Electrical Warning indi
instabilities of Fault Signal Transitions

WILLIAM COX

Contents

i

Foreword

Electric power systems are one of the central systems for services such as drinking water, telecommunications and banking. The day-to-day increase in the power demand and lack of simultaneous infrastructure development compel power systems to operate close to the stability margin. Power systems operating close to the stability margin are vulnerable to transitions leading to instability. The intermittent nature of renewable energy resources exacerbates the instability issues. The power generation with renewable resources varies as a function of time, imparting non-autonomous nature to power systems. The stability analysis performed on the power systems considers the system as autonomous. However, the stability regimes of autonomous and non-autonomous systems are different. This creates the need for a non-autonomous power system model which is simple enough to investigate the influence of rate-dependent variation of system parameters on the stability characteristics.

The power system is modelled as a non-autonomous system in the current thesis. Further, the stability characteristics of a non-autonomous power system model are investigated. Further, a comparative study of the stability regimes of autonomous and non-autonomous power system models is presented. The occurrence of subcritical Hopf bifurcation in the power system model was observed for the quasi-static variation of mechanical power. The rate-dependent variation in mechanical power showed a delay in the transition in the power system model. Next, depending on the rate of variation in mechanical power, early transitions with regard to the quasi-static Hopf point are observed. Furthermore, the relationship between the critical rate and the initial conditions is established.

Even though the system is quasi-statically stable, systems that are undergoing sub-critical Hopf bifurcation can nevertheless be triggered to the other oscillatory state by noise. Therefore, it is important to investigate the effect of stochasticity and intermittency on the stability regimes of power systems. Towards the same, the effect of noise on the transition characteristics is explored in the thesis. The stochastic changes from loads and generations in practical power systems are modelled as additive white Gaussian noise. Therefore, additive white Gaussian noise is adopted for our investigation as well. The delay in the transition observed for the variation of mechanical power as a function of time shows significant variability in the presence of noise. It is also noticed that if the angular velocity approaches the noise floor before crossing the unstable manifold, the rate at which the parameter evolves has no control over the transition characteristics. In such cases, the response of the system is purely controlled by the noise, and the system undergoes noise-induced transitions to limit cycle oscillations. The thesis uncovers the change in the width of the bi-stable zone for different intensities of noise in the system. Further, an emergency control technique to maintain the stable non-oscillatory state once the system has crossed the quasi-static bifurcation point is developed. The control technique allows electric utilities that operate near the physical limits to restrict operation within stable limits.

The oscillatory instability observed in power systems is sudden and catastrophic, which demands effective early warning measures. The thesis focuses on creating precursors for these unwanted transitions in the presence of noise. Critical slowing down-based early warning measures were developed for different rates of variation of system parameters and noise intensities. In order to determine how reliable these metrics are, the effectiveness of CSD-based early warning indicators was examined. In addition, a useful deep learning architecture for power system oscillatory instability prediction was put forth. Early warning systems based on deep learning algorithms are observed to be helpful in circumstances where CSD-based systems become unreliable.

The thesis demonstrates the possibility of reinstating the system to its initial non-oscillatory state, even after crossing a quasi-static bifurcation point. Early warning systems based on critical slowing down have been found inadequate for predicting oscillatory instability in power systems. Multiple realisations in the presence of noise were not showing the increasing trend in autocorrelation and standard deviation before the critical transition. Performance measures are used to verify the viability of the proposed deep learning method. By using non-autonomous modelling and stability analysis, combined with appropriate control techniques based on a mix of CSD-based EWS and deep learning architecture to predict imminent critical situations, increases the chances of predicting the occurrence of catastrophic transitions in power systems.

Chapter 1

Introduction

1.1 Background and Motivation

One of the key drivers in the pursuit of the progress of civilisation is the ability to harness energy. A series of agricultural and industrial revolutions facilitated the basic human necessities such as drinking water, heating and lighting homes, and fertilising and irrigation of crops to the requirements for the modern civilisation such as banking, telecommunications, and travel around the world [1]. The ability to identify and extract energy resources and convert the same into electrical power for more effortless operation accelerated the transformation of civilisation. Fossil fuels contribute to the major share of the overwhelming power consumption of the world. Burning of fossil fuels emits CO_x, SO_x, NO_x and volatile organic compounds into the atmosphere. These emissions increase the concentration of greenhouse gases and environmental pollutants in the atmosphere [1]. The natural process like plant photosynthesis regulates the concentration of CO_2 in the atmosphere by exchanging carbon between the atmosphere and the earth's land and oceans. However, the day by day increase in the anthropogenic carbon flux alters the balance that maintains the concentrations of CO_2 in the atmosphere. The increasing concentrations of greenhouse gases adversely affect the climate systems. The direct impact on the climate systems is evident from the rising frequencies of droughts, heatwaves, and torrential rainfalls followed by floods observed worldwide. These climatic impacts are existential threats to all-natural systems that support life [2]. Therefore, a drastic reduction in greenhouse gases is a prudent insurance policy rather than waiting for additional evidence. Hence, it is essential to have another 'green revolution' in which our energy sources are affordable, accessible and sustainable [3].

The UN-Paris agreement signed in 2015 put forward a long-term temperature goal of maintaining the global rise in the mean temperature well below 2^oC with respect to pre-industrial levels [4]. Phasing out of coal, oil and gas in the energy portfolio with renewable energy sources is essential to achieve this goal. The targeted and current development in the promotion of renewable energy is mainly focused on a few technologies due to the abundance of resources [5]. The most popular among those is wind energy, which is in service in various countries, including China, Denmark, Germany, India, the UK, and the US. The second popular renewable resource is solar energy due to its homogenous distribution. Other energy harvesting technologies from sources such as waves, tides, ocean currents, ocean thermal, and biomass are still relatively nascent. Despite the legal binding mechanism of the UN-Paris agreement, the energy production from fossil fuels increased from 80% to 81% over the years from 2000 to 2017 [2] to meet the increase in power demand. Therefore, there is a growing consensus for increasing the share of renewable energy sources in the energy portfolio to alleviate the impacts on the climate.

The increase in the penetration of renewable energy poses a paradigm shift in the nature of the energy, system dynamics and operational characteristics [6]. Regarding the nature of the resources, the output of the wind and solar generation varies depending upon the wind speed and available sunlight. This makes the renewable energy sources intermittent and distributed, unlike the conventional power plants [7]. This alters the generation architecture from the centralized generation (CG) to distributed generation (DG) [8]. The second major change is the change associated with system dynamics. The turbine and generator systems (for sources other than solar) that convert mechanical energy to electrical energy are often decoupled from the electrical grid in renewable energy systems. The solar power generation techniques do not involve any rotating parts [9]. Therefore, the overall inertia of the power system is significantly reduced with the augmentation of renewable energy systems. Further, the converter interfaced renewable generation introduces more power electronic devices into the power system. Additionally, the dynamic behaviour of the power electronic devices is faster compared to the time scales adopted for the analysis in traditional power systems [10]. The voltage sags resulting from the short-circuit faults interact with renewable energy sources, adding more uncertainty to the operating conditions [11]. Apart from all these, the harmonics associated with power electronic devices affect the power quality, which requires additional methods to improve the power quality [12]. In summary, even though renewable energy integration is essential in power systems, the paradigm shift in the system dynamics from that of the conventional power plants needs to be considered for the analysis.

The power imbalance in conventional power plants will be balanced by injecting or absorbing kinetic energy into or from the grid by the rotating masses to counteract the deviation in frequency [13]. The high inertia of the rotor of the synchronous machine limits the variation of the rotational frequency of the rotor in a conventional power plant. The reduction in the overall system inertia demands a dedicated penalizing mechanism to compensate for the fluctuations in frequency associated with renewable energy [11]. Further, power quality issues resulting from the harmonics necessitate additional circuitry to counter the same.

Snapshots of static analysis for a given load and generation scenarios are performed for the entire operating regime in conventional power systems to compute the stability margin [14]. This method of stability analysis is based on the assumption that the power system is autonomous. However, the feed-in fluctuations induced by the predominant renewable resources such as wind and solar are short term fluctuations with intermittent increment statistics [15]. This makes the renewable integrated power system a non-autonomous system. The effects of time-varying loads on the efficiency of the distributed generation are investigated and reported in the literature [16]. The proactive response of the prosumers necessitates the time-dependent variations of the backup generators as well [17]. Therefore, the time-varying nature of both generation and load is to be considered in order to avoid over or underestimating the benefits of distributed generation [7].

Power systems being a critical infrastructure essential for all services ranging from drinking water to telecommunication [18], tremendous effort is dedicated towards the maintenance and protection of the system [19]. The day to day increase in the power demand and lack of transmission line expansion compel the power systems to operate near the stability limits for the optimal usage of the assets [20]. Major power disturbances that occurred around the world in Canada and the US (2003), Germany, Belgium, Italy, France, Spain, the Netherlands (November 2006), and India (2012) are cascading failures that originated at remote locations

3

and propagated through the network [21]. These major blackouts testify that it is impossible to build insusceptible power grids. The intermittency and stochasticity associated with renewable energy resources have exacerbated the problem of instabilities [22]. Hence, developing reliable early warning indicators for power system instabilities is essential to ensure a secure and quality power supply. Therefore, this thesis focus on modelling the time-dependent variations occurring at the generation side, incorporating stochasticity of the renewable resources, and developing reliable early warning indicators for power system instabilities.

1.2 Mechanism and types of power system instabilities

Power system stability is much like the stability of any other dynamical system. The mechanism by which instability is initiated in a power system is detailed below to facilitate the process of identifying the apt method for power system stability. A fault occurring somewhere in the system disturbs the operating condition and triggers a sequence of events, including the functioning of the relay and circuit breaker to mitigate the fault. The system assumes a post-fault condition when the fault is cleared. However, the state of the system after clearing the fault may not be the desired equilibrium state [23]. System stability is the ability of the post-fault state to regain the desired equilibrium state. We can observe several definitions for power system stability with a rigorous mathematical background in the literature. Kundur *et al.* proposed the widely accepted definition of power system stability from the perspective of the physical system. According to Kundur *et al.*, "Power system stability can be defined as the ability of an electric power system, operating under a set of prefixed initial conditions to regain a state of operating equilibrium after being subjected to a physical disturbance, where most of the state variables are bounded so that practically, the entire system remains intact" [24]. Depending upon the magnitude and severity of the disturbances, power system stability can be classified as small-signal stability are large-signal stability. The small-signal stability and large-signal stability is also known as steady-state stability and transient stability respectively in power systems. Steady-state stability is defined as the ability to maintain steady-state equilibrium upon a small disturbance in the operating conditions. Transient stability of a power system is the ability of a system to maintain the equilibrium following a large disturbance such as a sudden change in load and generation, change in the transmission system configuration

4

due to faults and line switching [25].

The prevalent method to analyse the stability of the power system is by linearising the system dynamics about the operating point. The successive works analysed the stability of the linearised system using techniques such as Nyquist or Routh-Hurwitz stability criterion, sensitivity analysis and eigenvalue analysis [26]. The limited knowledge of the validity where the linearisation is accurate poses a challenge to the linearisation approach. This restricts the linear stability analysis to small-signal perturbations occurring in the power systems. Therefore, in the earlier stages of power system development, transient instability is considered as a crucial challenge for the power system stability. More effort and research were dedicated to analysing the transient stability problem occurring in power systems. These researches identified the system parameters that violate the operational bounds during the transient stability problems in power systems.

One of the system parameters which is affected during a transient instability is voltage, which is known as voltage instability. Voltage instability is a severe power system stability issue exhibited by interconnected power systems characterized by low/ unacceptable voltage at the system buses [27]. For power systems operating at highly stressed conditions, voltage instability further leads to a situation of the progressive decline of the voltage at the system buses, referred to as voltage collapse [28]. There are incidents of voltage collapse leading to blackouts in history, such as in France (Dec 1978) and in Belgium (Aug 1981) [29]. A second vulnerable system parameter in response to disturbances of severe magnitudes is the system frequency. In frequency instability, the system is not able to maintain steady-state frequency following a severe imbalance in load and generation [26]. Another major stability problem that gained significant research interest is rotor angle stability which is the ability of interconnected synchronous machines to remain in synchronism [23]. The rotor angle stability problem involves the study of electromechanical oscillations inherent in power systems.

The oscillatory instability of interconnected systems aggravated with the spread of the electrical network. These are electromechanical oscillations of low amplitude frequencies, observed in large interconnected systems connected by relatively weaker tie-lines [30]. Depending on the number of generators involved in these oscillations, these are classified as local mode and inter-area oscillations. The oscillations associated with a single generator or plant are local

5

modes. Generators are also observed to form groups and oscillate against each other, with frequencies ranging from 0.1 to 0.8 Hz. These oscillations associated with groups of generators are known as inter-area oscillations [31]. Additionally, the high-frequency switching operation of the power electronic devices induces harmonic instability which is absent in conventional power systems [32]. These harmonic oscillations can trigger series and parallel resonance with frequencies of hundreds of Hz to several kHz.

1.3   Power system stability analysis methods

A power system is strongly nonlinear due to the nonlinear nature of the governing equation and the loads. However, the initial attempts to analyse the stability of the system were based on linearisation methods. Gao *et al.* proposed a 'modal analysis technique' to analyse the stability of the power system [33] based on the mathematical model of the system. The authors derived a set of smallest eigenvalues and the associated eigenvectors from the reduced Jacobian matrix of the steady-state system. The reduced Jacobian matrix is computed for variations in the reactive power while maintaining the $Q - V$ relationship of the network.

In [34], Ajjarappu and Christy proposed a method to find the continuum of power flow solutions starting at some base load, leading to steady-state stability limits. In this method, the authors presented a continuum of power flow solutions, starting at some baseload leading to the steady-state stability limit of the system. A predictor-corrector scheme is employed to find a solution path to the power flow equations, for variations in the loading parameter. Here, the solution space starts from a known solution and proceeds to estimate the subsequent solution using a tangent predictor. The tangent vector provides information about the differential change in the voltage bus for a differential change in the load. This information is used to identify the weakest bus and formulate the sensitivity index. The left and right eigenvectors of the Jacobian matrix for increments in loading parameters are computed in the point of collapse (PoC) method used for evaluating the loadability margin. A comparison of PoC and continuation methods for computation of voltage collapse in power systems is presented in [35].

Further, Morison *et al.* [36] presented a comparative voltage stability analysis using static and dynamic methods. The static analysis is carried out using the 'modal analysis technique' for different operating conditions of the system along the time-domain trajectory. Time-domain simulations are performed, modelling various network components that impact the stability, such as loads, underload tap changers (ULTC), and maximum excitation limiters (MXL) of the generators in dynamic analysis. The study concluded the consistency of the stability regimes predicted using static and dynamic analysis for the operating regime considered. Nonetheless, these methods were insufficient to provide a realistic picture of the power systems, especially when operating region is bi-stable. This sparked research interest in developing methods that are suitable for power systems.

The initial approaches towards this goal were based on time-domain simulations. Time-domain simulations are performed by mathematically modelling the system in a simulator. Further, the system behaviour for different scenarios of disturbances and parametric regimes is simulated. The result of the complete simulation gives information about the stability margin of the system. However, the detailed information about the points of the swing equation of the system is not necessary to identify the stability margins of the system. Therefore, research interest shifted to developing methods that use only partial or no use of the differential equation to compute the stability regimes faster.

Aylett *et al.* proposed a new method to study the limits of the transient stability of power systems using energy integral and singular points, which require only the partial solution of the nonlinear swing equations [37]. The magnitudes of kinetic and potential energy along the separatrix are equated to formulate the energy equation. The transient stability of the post-fault system is evaluated by substituting the values of the post-fault system variables in the energy equation. The system whose post fault energy lesser than the energy of separatrix is identified as stable. The determination of stability using energy integral initiated the formalism of Lyapunov methods for evaluating the transient stability of power systems. The Lyapunov methods estimated the region of asymptotic stability of the equilibrium point and the critical switching time without solving the differential equations associated with a disturbance [38], [39]. Further studies in this area focused on the development of the Lyapunov energy function by applying the Popov Criterion, which is custom-made for nonlinear systems [40]. The correctness of the mathematical model, which is often a challenge, adversely affected the accuracy of the Lyapunov energy method. In order to address this problem, the asymptotic stability around an equilibrium point or the basin of attraction of the equilibrium point were analysed using structure-preserving energy functions [41]. Recently, studies were reported on assessing the global stability by determining the largest Lyapunov exponents [42].

Further, Kakimoto et al. proposed a potential energy boundary surface (PEBS) method for the estimation of stability based on lure type Lyapunov function [43]. In this method, the closest unstable equilibrium point [25] that possesses the lowest value of energy among all unstable equilibrium points is evaluated on the stability boundary, and a conservative estimate of the stability is obtained. However, the computation of the closest unstable equilibrium point (u.e.p) is independent of fault location. Furthermore, Athay et al. presented an analytical method based on energy function that retains the transfer conductance of the system with respect to the fault location for the direct determination of transient stability [44]. The controlling unstable equilibrium point is a viable method among the direct techniques for transient stability. However, the issue in controlling unstable equilibrium points is determining the precise controlling unstable equilibrium points relative to fault-on-trajectory. Even though the energy function method based on controlling unstable equilibrium point got acceptance, the practical application is hindered by model complexity and computational burden.

In [45], the authors developed a boundary of stability criterion based controlling unstable (BCU) equilibrium point of the power system from a reduced-order system. The computation of the controlling unstable equilibrium point of the reduced-order system is easier and computationally cheaper. The efficacy of the method is evaluated on standard power system models with different numerical schemes to find the exact controlling unstable equilibrium point relative to fault-on-trajectory. Llamas et al. reported a conflicting observation wherein the power system goes to instability, even though the equilibrium point associated with the first swing is stable [46]. The observation raised scepticism about the reliability of the stability margins evaluated via BCU approach.

Further, the voltage stability analysis tools combining both static and dynamic approaches, which use techniques such as time-domain simulations, modal analysis and continuation power flow, were developed. Voltage Stability Analysis Program (VSTAB) is one of the tools based on modal analysis and continuation power flow to assess the static stability, which gained significant research interest. Continuation power flow is used to determine voltage stability margin (VSM) and the bus voltages at the point of collapse. Modal analysis is used concurrently to determine the critical modes and their associate buses and to identify voltage instability characteristics.

In summary, the literature on the static instability mechanism of power systems illustrated hitherto is based on linearizing the system dynamics. The dynamic stability analysis is centred around the Lyapunov based energy function approach and time-domain simulations. The main drawback of the former is the difficulty in determining the apt Lyapunov function, which depends upon the accuracy of the model. Several investigations on power systems revealed the co-existence of multiple stable states [47]. The lack of information about the basin of the attraction of the equilibrium point restricts the application of time-domain simulations that depends upon the initial conditions of the system. Therefore, understanding the basins of attraction of the equilibrium point is necessary to estimate the stability margin precisely. With this view, stability from the perspective of bifurcation theory is investigated to understand the suitability of this technique in power systems.

1.4   Power system stability analysis from the perspective of bifurcation theory

Even though the power system is known to be nonlinear, researchers often use linearisation-based stability analysis and tools for the assessment. In the early 90s, researchers explained the significant blackouts characterised by the progressive decline of voltage at the system buses from the perspective of nonlinear analysis and bifurcation theory. Dobson and Chiang elucidated the loss of stability from the perspective of bifurcation theory for the first time. The authors observed the disappearance of a stable equilibrium point via a saddle-node bifurcation in a simple power system model [28]. Here, the authors demonstrated the movement of the system state along the centre manifold of order one, culminating in a saddle-node bifurcation in a simple power system model.

The study by Dobson *et al.* and subsequent investigations analysed the loss of stability by vanishing of an equilibrium point for the variations in the control parameter [28], [48]. Another major instability is oscillatory instability, characterised by low amplitude sustained oscillations, resulting from pole slippage of synchronous machines [49]. These oscillations are

9

known as limit cycle oscillations. These oscillations are strong enough to damage the costly equipment over the years of operation. The relays employed in power systems are designed not to interfere with oscillations during the transient period. These relays cannot distinguish sustained oscillations from those occurring during the transient period. Hence, knowledge about the system parameters is essential to restrict the operation within the stable region. With this rationale, the study of bifurcations from a stationary point to a family of periodic orbits of dimensions greater than or equal to two was investigated in power systems. Abed and Varaiya identified the loss of stability of equilibrium point by interacting with a limit cycle via a Hopf bifurcation [50]. This investigation revealed the possibility of a loss of stability prior to the point of saddle-node bifurcation, which was neglected in the previous investigation [28]. The authors discovered that power systems undergo Hopf bifurcation to periodic orbits for variations due to (1) variable net damping, (2) frequency dependence of electrical torque, (3) lossy transmission line and (4) excitation control.

Subsequently, Ajjarapu and Lee established the existence of a stable and unstable periodic oscillation and voltage collapse in a power system model for variations in reactive power load before saddle-node bifurcation [47]. The stability of the periodic solutions is analysed by determining the Floquet multipliers of the monodromy matrix. A successive study by Abed *et al.* closely investigated the influence of Hopf and period-doubling bifurcations on the stability margin and its role in the phenomena of voltage collapse [51], employing the same power system model as in [28]. The authors identified the chances of loss of stability of the nominal equilibrium point through cyclic fold bifurcations and Hopf bifurcations for variations in reactive power demand, leading to a reduction in the stability margin.

Later, Venkatasubramanian et al. observed the possibility of a singular limit for a system with an infinitely fast time scale and the need for modelling the system as a differential-algebraic equation (DAE) to handle the challenge. [52]. The changes in the eigenstructure of the system Jacobian and the occurrence of singularity induced bifurcations (SIB) are mapped in the study. The loss of stability of the feasible boundary occurs when the equilibrium point is at the singular surface. Afterwards, Marszalek and Trzaska conducted further research on SIB in DAE and their occurrence in power systems [53]. The authors detected the occurrence of single and double singular points, which can be identified from the corresponding change in the index of the matrix pencil.

The bifurcation analysis covered hitherto in power systems studied the quasi-static variation of the control parameter, assuming the system as autonomous. However, the inherent nature of the prominent renewable sources such as wind and solar power is continuous variations and intermittency [7]. This makes renewable energy integrated modern power system a non-autonomous system. Therefore, the non-autonomous variations occurring at both the generation and load sides are to be considered while performing bifurcation analysis, which is explained in the next section.

1.5   Need for power system stability analysis preserving the non-autonomous nature of the system

Rate-dependent bifurcations, where the bifurcation parameter varies as a function of time, are available in the literature using canonical models of standard bifurcations. Baer *et al.* reported the the phenomena of delay or memory effect for slow passage through Hopf bifurcation point for the first time [54]. Baer *et al.* observed a delay in the transition from a stable non-oscillatory state to the oscillatory state for the variation of system parameters as a function of time in a relaxation oscillator [54]. Additionally, the authors observed that the presence of small amplitude noise and periodic perturbations of near-resonant frequency destroys the delay or memory effect. Successively, Majumdar *et al.* established the delay in transition to the alternate state for the slow variation of the control parameter and the dependence of the delay upon initial conditions in nematic liquid crystals [55].

Dynamical systems exhibit periodic solutions associated with Hopf bifurcations point. Many of these oscillations are undesirable. Therefore, the delay effect for slow passage through Hopf bifurcation gained significant research interest from the dynamical system community. In [56], Ashwin *et al.* observed a sudden transition from a stable state to an alternate stable state due to the rate of variation of the bifurcation parameter. This novel mechanism of triggering transitions in dynamical systems where multiple stable states co-exist is termed rate-induced tipping (R-tipping). Additionally, the authors noticed that R-tipping mechanisms act independently of the presence or absence of other types of transition mechanisms such as bifurcation (B-tipping) and noise (N-tipping). Their study on climate systems reveals an entirely new mechanism arising from rate and the possibility of suppression of B-tipping for fast variations of system parameters. Soon, the possibility of R-tipping is investigated in other dynamical systems such as biological systems [57], climate systems [58], and chemical systems [59].

Subsequently, Tony *et.al* performed the rate-dependent variation of system parameters in a thermoacoustic system and investigated the transitions induced solely due to the rate-dependent variation of system parameters within the bistable region [60]. The studies on the aforesaid dynamical systems established that there are chances of transition purely due to rate with the presence or absence of bifurcation and noise. In summary, it is concluded that the stability regimes of the autonomous and non-autonomous systems are different.

The stability analysis methods reported in the literature of power systems do not consider the non-autonomous nature of the physical system. Nonetheless, evidence of a difference in stability regimes of autonomous and non-autonomous systems suggests that it is appropriate to analyze the stability of the power system, preserving the non-autonomous nature. In physical systems, transition points may be a combination of all three mechanisms such as bifurcation, noise and rate [61]. From the observations mentioned earlier, the objectives of the thesis are formulated.

1.6   Effect of noise in dynamical systems

A dynamical system that undergoes subcritical transition will remain in the non-oscillatory state as long as the perturbations are within the basin of attraction of a fixed point. The system will start oscillating when the magnitude of perturbations crosses the basin boundary of the equilibrium point and move towards the basin of attraction of the limit cycle. In physical systems, uncertainty is inevitable due to measuring, material, manufacturing, and assembling. These uncertainties can be modelled as random parameters with specific statistics. In recent years, there has been significant research interest in studying nonlinear dynamical systems with random perturbed parameters due to their practical application.

Ma *et al.* investigated the effect of bounded random parameter subjected to harmonic excitation by Chebyshev polynomial approximation [62]. The authors observed bifurcation for variations in amplitude, frequency and intensity of random perturbations in the stochastic system. Further, the authors identified stochastic period-doubling bifurcation due to random parameter variation alone. Thereafter, Li *et al.* examined the chaotic behaviour of the van der Pol-Mathieu-Duffing oscillator under bounded noise by random Melnikov method [63]. The mean square criterion is used to detect the necessary conditions to establish the chaotic motion of a stochastic system. The results show that bounded noise can enhance the largest Lyapunov exponent.

In [64], authors establish that uncorrupted basins are necessary for safe, practical application, and eroded basins constitute a critical state for the structure leading to failure. Even though basin erosion can be prevented using structural modification, the approach is conservative as the system can survive the erosion if the erosion is not sharp. Therefore, it is necessary to investigate the safe basin erosion in dynamic systems. The authors examined how the amplitude of external excitation influences the basin erosion in Helmholtz, Duffing, and rigid block oscillator models in [65]. A rapid basin erosion in ϕ^6 van der Pol oscillator triggered by the amplitude of parametric excitation is reported in [66]. It is identified that random noise excitation plays a vital role in the basin erosion of nonlinear dynamical systems.

In [67] Gan examined the effect of noise on the erosion of safe basin and noise-induced chaos in softening Duffing oscillator using Melnikov integral. The author observed random intersection of the stable and unstable manifolds and the presence of chaotic motion in the absence of damping in the noisy system. Further, basin erosion is observed in the stochastic system, which is aggravated for increase in the noise intensity.

1.6.1 Effect of noise in the dynamics of power systems

The integration of renewable energy sources and plug in-electric vehicles (PEV) introduces a significant amount of uncertainty in the power system operation due to the stochasticity associated with these elements. The uncertainty in the power system affects time scales ranging from automatic generation control to day-ahead scheduling. Due to the stochasticity, there are variations in the stability boundaries of the stochastic and deterministic systems. The uncertainties in the power system are originated due to (i) randomness of the initial equilibrium point resulting from stochastic disturbance and fluctuations of power flow, (ii) randomness of the state matrix co-efficient resulting from variations of system parameters of the network, (iii) stochastic external excitation such wind turbine mechanical input [68].

Considering the small perturbations in the operating characteristics of transmission lines and the loads, stochastic modelling of power systems has been of research interest since the early 1990s. The perturbations in the power systems are modelled by a Gaussian stationary process (white) with constant spectral density [69]. The authors consider the perturbations occurring in the load to model the stochastic power system. The authors evaluated the elapsed time before crossing the stability boundary, termed as mean first passage time, as the performance measure. Successively Wei *et al.* studied the effect of Gaussian white noise on basin erosion in a simple power system model [70]. The authors employed stochastic Melnikov method to investigate basin erosion. The authors observed that the safe basin erodes after a threshold noise intensity, and the problem is exacerbated when the noise intensity is greater than the threshold.

In [71], the authors investigated the influence of random parameters on the dynamical behaviours of a single machine infinite bus(SMIB) power system model. Initially, the deterministic power system model is allowed to operate in the parameter range where the system exhibits periodic oscillations. The effect of intensity of the random parameter on the power system is assessed by the log of power spectra, leading Lyapunov exponent, and phase diagram. The authors observed that the power system model falls into chaos with increased noise intensity.

Chen *et al.* studied chaotic motions and subharmonic bifurcations in a nonlinear power system [72]. The periodical perturbations due to errors or machinery rotation in the generators are considered here. The authors applied the three-parameter Melnikov method to the swing equation of the power system for the analysis. Initially, the conditions for heteroclinic bifurcations and chaotic behaviours are obtained by computing the energy of the non-perturbed, conservative system. Further, the conditions for heteroclinic bifurcations in the perturbated, dissipative system is identified . The results illustrated the presence of heteroclinic bifurcations for variations in damping coefficient, machinery power, and intensity of perturbations in the stochastic operating regime.

Odun Ayo *et al.* proposed a stochastic energy function on a structure preserved power system model and presented a probability metric associated with a predefined randomness to assess the transient stability [73]. The classical transient stability studies on an equivalent reduced-order model of the power system network was used to formulate the energy function. However, the reduced-order method does not address the variation in the reactive power demand and the voltage at the system buses. As an alternative, the authors proposed a structure preserved power system model that maintains the structural integrity same as the original power system model. Here, an appropriately scaled Weiner process represents the power system frequency deviations. Monte Carlo simulations were performed to ascertain the system behaviour over multiple trials to address the randomness. The effect of randomness is reflected in the variations in the critical clearing time over multiple trials. The critical clearing time of the deterministic case and the mean value of multiple trials was observed as constant. However, the probabilistic analysis reveals the risk associated with the randomness.

Dhople *et al.* proposed a framework of the stochastic hybrid system to analyse the small-signal stability in the context of RES and PEV [74]. The power system dynamics are represented using differential-algebraic equations (DAE). The active/reactive power injections are modelled using a continuous-time Markov chain for accounting uncertainties of the RES. The system dynamics are linearised around the operating point to evaluate the small-signal stability and to analyse the impact of stochastic injections. Further, the statistics of the dynamical states are computed using an analytical method formulated using Dynkin's formula.

In [68], Yuan *et al.* investigated the small-signal stability of power systems for stochastic external excitation. The stochastic variations of the mechanical power input of the wind turbine generator are represented by a Weiner process. It was observed that even for negative eigenvalues of the system state matrix, stochastic excitation could lead to bounded oscillations in the power system. In [75], Zhou *et al.* analysed the chaotic dynamics resulting from a periodic disturbance in a SMIB power system model with Vanderpol damping. The authors computed the Melnikov function associated with the pair of heteroclinic orbits and the conditions for chaotic motions. Further, the critical curves separating the chaotic and non-chaotic regions were presented. Small signal stability of power systems with uncertainties is examined in [76]. The authors modelled the stochastic continuous variations by a Wiener process in the mathematical model of the power system. Further, the stability of the linearised system is assessed. The proposed probabilistic method can assess stability by evaluating the algebraic expression without solving the partial differential equations.

The literature on the stability of the power system reveals that it is possible to identify safe operating regimes using the existing methods. However, demand-side management implemented in the modern power system necessitates an early warning of the instabilities for taking alternative measures. The precursors for instabilities in the power system literature are presented in the coming section.

1.7 Precursors for transition in power systems

The sudden large-amplitude oscillations at the point of instability in a bi-stable region are characteristic of subcritical transitions. The irreversible nature of these transitions makes subcritical transitions crucial and challenging [77], [78]. The power system also exhibits limit-cycle oscillations, which are detrimental to the system in the long run. Therefore, it is necessary to develop precursors for the imminent catastrophic transitions in power systems.

A salient class of precursors for imminent regime shifts based on critical slowing down (CSD) gained significant popularity among dynamical systems. The real part of the dominant eigenvalue attains values closer to zero while the system undergoes a bifurcation. As a result, the time taken to die out the perturbations that the system undergoes increases near the bifurcation point. This phenomenon is referred to as critical slowing down. The slow decay of the perturbations associated with critical slowing down signals shows an increase in the statistical measures such as autocorrelation and variance.

The complex nature of the physical systems makes it difficult to build models that describe their dynamics accurately. This accelerated the development of critical slowing down-based precursors from the time series data of dynamical systems. Dakos *et al.* discovered that the dynamical systems show critical slowing down inherent to critical points irrespective of the difference in dynamics of the systems, as an early warning before transition [79]. They also noticed that the dynamical system undergoes rapid changes driven by internal feedback upon crossing a critical threshold.

Scheffer *et al.* proposed several statistical measures such as recovery rate, variance, and autocorrelation derived from the phenomena of critical slowing down to alarm catastrophic regime shifts in ecosystems [77]. Subsequently, these early warning signals are universally applied in medicine [80], finance [81], power systems [82] and other engineering systems [83]. Sheffer *et al.* explained the principle behind the functioning of early warning indicators in a sequel work [84]. Scheffer et al. also formulated early warning indicators based on flickering, wherein the nonlinear systems switch between the alternate stable states in a bi-stable system in the presence of noise.

Cotilla *et al.* identified critical slowing down based early warning signals in the real-time data of 10, Aug 1996 blackout in power systems. They proposed autocorrelation and variance from a single stream of synchronised phasor measurements using CSD to predict impending instabilities in power systems. The authors also modelled the power system as a stochastic slow-fast system to derive empirical results for the critical transitions exhibited by power systems.

Even though a multitude of results reported the efficacy of critical slowing down based early warning signals [79], [77], exceptional results which mark the silent transition to the alternate state and partial failure of these measures are also reported [85], [86]. Additionally, Ke'fi *et al.* analysed the behaviour of slowing down based early warning indicators pertaining to catastrophic and non-catastrophic situations [87]. The authors identified that decrease in recovery rate is not a characteristic of an imminent catastrophic shift. Instead, it is a characteristic of being increasingly sensitive to perturbations, independent of whether the impending transition is catastrophic or not. This study confirmed the chances of false-positive alarms in slowing down based early warning indicators.

The early warning indicators of regime shifts were proposed based on the physical notion of potential energy, which guides the expected dynamics. Empirical validation of the leading indicators in the presence of a potential that depends smoothly on the underlying state of the system and changes smoothly with respect to underlying conditions gave acceptance for these indicators. However, the pioneering work by Hastings *et al.* [85] reported the absence of smooth potentials prior to regime shifts in ecological systems, which undergo slowly varying operating conditions via numerical examination. Their studies offer a cautionary note to the usage of critical slowing down based early warning signals. Subsequently, Boerlijst *et al.* [86] argued that transitions can occur without an early warning with the help of three counter examples of silent catastrophes in accepted ecological models. The authors also discovered that silent catastrophes occur when the perturbations are not in the direction of the dominant eigenvector across which the system destabilises, leading to the absence of generic early warning signals. Thus, the authors conclude that characteristics of slowing down may not be visible in all the variables.

Dakos *et al.* proposed a method combining 'metric-based indicators' and 'model-based indicators' to identify impending critical transition in two simulated data sets. The authors observed that a single indicator or method is not robust, in line with the previous observations [86], [85]. Further, the study stressed the need for specific data treatment to procure sensible information.

Ghanavati *et al.* investigated the robustness of critical slowing down based early warning indicators to predict the critical transitions that occur in power systems [82]. The work was inspired by the observation drawn from ecological systems wherein the critical slowing down based early warning indicators do not appear early enough for an effective early warning indicator in all the variables. Ghanavtai *et al.* noticed that even though slowing down is signalled before the transition in power systems, the increase in autocorrelation is reflected in a minimal set of variables. There will not be any early warning signals if the variables that are being monitored are different. Further, a sequel study from the same group of authors proposed an analytical method to indicate proximity to a bifurcation [20]. After that, Zheng *et al.* proposed a critical transition framework based on the dominance of the external stochastic fluctuations for voltage collapse prediction [88]. The previous studies observed that critical slowing down based early warning indicators are not always present in all the variables. The authors proposed a semi-analytical solution for variance based on the linearisation of the system about the equilibrium point. Further, a scaling law was also established between the variance and the distance to collapse.

Wissenfield proposed the pioneering study on developing precursors to predict imminent transitions associated with co-dimension one bifurcation in the presence of an optimum amount of noise [89]. Early identification of the impending bifurcation and its nature is predicted here by observing the width of the peak frequency of the amplitude spectrum. However, the linearisation of the perturbation equation poses a limitation to its applicability in highly nonlinear physical systems.

18

Shalalfeh *et al.* proposed an alternative method to predict the distance to instability by monitoring the fractal dimensions of the time series. Initially, the non-stationarity of the data is proved using unit root tests, Augmented Dicky- Fuller (ADF) and Kwiatkowski-Philips-Schmidt-Shin (KPSS) [90]. Further, the fractal parameters which exhibit long-range correlation in the data of voltage magnitude, frequency, and phase angle are computed by detrended fluctuation analysis(DFA). Then, to develop early warning signals, Kendall's tau of the Hurst exponent is calculated for the frequency data under normal operation and before the collapse. The autoregressive fractionally integrated moving average (ARFIMA) model is found apt for the PMU data. The difficulty in extracting useful information from the copious PMU data is challenging. This hinders the applicability of these models for real-time applications.

A significant breakthrough has been made in artificial intelligence-based data-driven techniques that demonstrate promising results for online monitoring and security of physical systems. The wide deployment of a phasor measurement units and accelerated growth of artificial intelligence techniques shifted the research interest to data-driven machine learning methods such as artificial neural networks, decision trees, support vector machines, K nearest neighbours, and extreme learning machines. The initial step for implementing intelligent data analytics is to train the intelligent model using simulated / measurement data from the historical operating sequences and the labelled security status. Further, the incoming phasor measurement units are fed to the trained model to formulate online dynamic security assessment decisions [91].

Sun *et al.* proposed a decision tree algorithm for the online security assessment of power system, wherein the data from the phasor measurement unit is utilized [92]. In the decision tree algorithm, the threshold value that results in insecurity in the dynamic performance is identified. The critical attributes from the system parameters are computed for every contingency offline. The real-time data from the phasor measurement units are compared with critical attributes stored in the decision tree to determine the terminal nodes and the associated paths in the event of contingencies. Further, an insecurity score is computed for each path, and an alarm signal is generated for scores that exceed the pre-set limit. The recent success of the artificial intelligence-based dynamic security assessment methods assumes that phasor measurement unit data collected at the control centre is complete. However, the phasor measurement unit provides only partial data in events of communication loss, phasor data concentrator failure, and phasor measurement unit malfunction. This significantly impairs the decision making of the artificial intelligence-based techniques. Surrogate splits [93] and random subspace [94] based decision tree ensemble architecture are used to maintain the validity of a decision tree under incomplete measurements. Nonetheless, these approaches are black-box models due to the stochasticity of the training mechanism. Online dynamic security assessment in the event of incomplete phasor measurement units with the help of a robust white box decision tree ensemble architecture is proposed in [95].

A nonlinear classifier based on support vector machines for the online transient stability assessment of power systems is presented in [96]. A support vector machine is a linear machine in a high dimensional feature space, related to the input space in a nonlinear fashion, which allows the development of faster training techniques. The feature selection capabilities in higher dimensional spaces make support vector machines a suitable choice for the prediction in bulk power systems. Moulin *et al.* presented a comparison of the multi-layer perceptron and the support vector machine for the online transient stability assessment of the power system [97]. However, machine learning algorithms require careful feature selection or the representation of the data to make correct decisions. Post contingency stability assessment in the event of transient stability is carried out online using long-short-term memory(LSTM) to extract features from the phasor measurement unit data in [98]. Here, temporal data dependencies are utilized to extract features for better assessment accuracy in the model, which is different from the existing machine learning implementations in power systems literature. Deep learning solves the problem by feeding the raw data to the machine, which automatically detects the representation needed for the classification.

The transient stability assessment and classification of the aperiodic and oscillatory instability modes is performed in [99]. A deep neural network consisting of a convolutional layer and a fully connected layer is employed for the stability assessment in this work. Comparison of the optimization algorithms such as stochastic gradient descent and stochastic gradient descent with a warm restart is considered during the offline training here. A nonlinear dimensionality reduction technique, termed t-distributed stochastic neighbour embedding, is used for effective feature extraction.

1.8 Summary of the state-of-the-art literature on the non-autonomous power system model and early necessity of early warning indicators

In summary, the parameters of the power system such as generation, load, and inertia vary as a function of time [100]. Nonetheless, the non-autonomous variation of the system parameters is not covered in the literature on the power system to the best of the author's knowledge. Further, the transition from non-oscillatory state to oscillatory state is subcritical in power system [47]. The presence of delay or advance effect due to slow variation of system parameter through the bifurcation point is observed in other dynamical systems [54]. However, the effect of the non-autonomous variation of the system parameters on the transition characteristics of power systems is unexplored, which is crucial when the system is undergoing a subcritical transition.

The existing literature on the effect of noise in power focuses on the reduction in the safe basin with noise. However, the effect of the noise on the bi-stable characteristics of the power system model is unexplored. The phenomena of noise-induced triggering, wherein the system switches from a stable state to an alternate stable state within the bi-stable region when the noise intensity has crossed a threshold value, has not been investigated in power systems.

Even though precursors based on slowing down are applied to predict imminent catastrophic transitions in power systems, the robustness of these measures is not investigated. Stochasticity is identified as the signature feature behind the functioning of critical slowing down based early warning indicators. Hence, there are chances that multiple realisations can lead to different results. The efficacy of artificial intelligence-based techniques in predicting the transitions in dynamic systems is well established. Therefore, it is worth developing an artificial intelligence-based predictive mechanism to detect the sudden transitions occurring in non-autonomous power system models, which is easier for real-time implementation with appropriate scaling.

1.9 Objectives of the thesis:

The objective of the present work is to understand the influence of time-dependent variation of system parameters and the effect of noise on the stability characteristics of a power system model. Further, the study aims to develop early warning measures for CT observed in the power systems. The primary objectives are:

1. To understand the effect of variation of system parameters as a function of time on the transition characteristics of a power system model.

2. To investigate the effects of external noise on the bi-stable characteristics and transition characteristics of a power system model.

3. To develop reliable early warning measures of critical transitions observed in power systems by comparing critical slowing down (CSD) based early warning signals (EWS) and deep learning-based EWS.

1.10 Overview of the thesis

The objective of this study was achieved with the help of numerical simulations performed on a mathematical model of a single machine connected to an infinite bus (SMIB). The time series procured from the mathematical model are analyzed using the tools from dynamical system theory. The numerical model is perturbed with additive white Gaussian noise to understand the effects of noise on the power system's bi-stable characteristics and transition characteristics. Early warning measures were developed by exploiting the concepts of CSD. Further, the efficacy of CSD based EWS is determined, and a deep learning-based EWS is proposed.

The rest of the thesis is organised as follows. Chapter 2 introduces the mathematical model adopted for the study and the rationale for selecting the same. Further, the Chapter describes the tools that we have used to carry out this work. This Chapter also includes the bifurcations in power systems and the associated normal form equations. Finally, the development of CSD based early warning signals to predict critical transitions for dynamical systems is also presented.

Chapter 3 describes the influence of the rate-dependent variation of system parameters on the transition characteristics of a power system model. This chapter also explains the correlation between the initial conditions and the rate of variation of system parameters in determining stability. Further, the early transition to the alternate and the reduction in the basin of attraction with respect to initial conditions is also illustrated. The observations are published in the article "Rate-induced transitions and advanced take-off in power systems".

The reduction in the width of the bi-stable region with an increase in noise intensity is discussed in Chapter 4. The complete suppression of the bi-stable region in the presence of high-intensity noise is also covered in this Chapter. The interplay between the noise and the rate of the transition characteristics on a power system model is explained finally. The observations are published in the article "Emergency rate-driven control for rotor-angle instability in power systems".

Chapter 5 illustrates the development of early warning measures for predicting imminent transitions observed in power systems. The concept of critical slowing down (CSD) near a bifurcation is exploited to develop early warning measures. The measures such as lag-1 auto-correlation, variance, and return rate are adopted for the formulation of early warning signals. Then, the robustness of these measures in predicting a transition is investigated. Further, an artificial intelligence-based prediction architecture to forecast impending critical transitions in power systems is proposed. Finally, the performance of the artificial intelligence-based EWS with CSD based EWS is compared in this Chapter. A manuscript is in preparation based on the results for submission to PeerJ Computer Science.

The conclusions that can be drawn from the present study are elaborated in Chapter 6. This Chapter also proposes a possible extension of the present work on complex power system models to pursue in future.

Chapter 2

Background on the power system model and methodology

This chapter presents a detailed description of the canonical power system model adopted for the investigation in this work. Further, a background on the various types of bifurcations observed in power systems is illustrated. Thereafter, the procedure for developing a numerical bifurcation diagram is demonstrated with the help of a normal form equation. The formulation of early warning indicators based on the critical slowing down exhibited by the data is delineated in the last section.

2.1 Power system model

The intrinsic property of high nonlinearity makes the electric power system complex. Studying the dynamics of power systems is essential to ensure rotor-angle stability, voltage stability, frequency stability and a reliable power system. Apart from the three instability mechanisms, power quality issues also affect the power systems. However, the former is crucial to the operational dynamics. In order to facilitate the study of power system dynamics, power system models are adopted.

There are different modelling choices that facilitate the stability analysis of power systems. The first model among the approaches assumes a reduced order model, where the entire network is reduced to an n-port, as seen from the n-generator internal buses [41]. In this approach, the swing equation models the generator, and the load is assumed to be of constant impedance, where the structure of the original network is lost. This approach has a few shortcomings, such as neglecting variations in reactive power demand and voltage variations. Bergen and Hill proposed an alternate method to analyse the stability of the power system via structure-preserving models [101]. The structure-preserving model explicitly represents the active and reactive power demand at each load bus. The benchmark systems adopted for analysing and controlling electromechanical oscillations in power systems are explained below.

Single machine infinite bus(SMIB) test systems have extensively been used to study electromechanical oscillations. In SMIB, all the generating sources are lumped at one generation point, which feeds to a bus of infinite capacity. SMIB power system model is found effective in investigating the practical aspects of field commissioning and testing of stabilizers [102]. Nonetheless, the simplification inherent to the SMIB model limits its applicability to study inter-area oscillations seen in large interconnected systems. In order to investigate these aspects, multi-machine power system models were developed.

Small-scale systems that retain the characteristics of interest, such as inter-area oscillations, are chosen for the study of multi-machine power system studies. The simplest of those systems is the 3 machine infinite bus power system which consists of 3 generators and 6 buses [102]. In this model, generators 1 and 2 are identical and are connected in parallel. 3MIB model exhibits intra-plant, inter-plant and inter-area oscillations. A seven bus five machine equivalent model of the south-southeastern Brazilian system configuration was developed in the 1990s [103]. Generator7 in this model is equivalent to the Southeastern area of Brazil. Rigorous investigations were performed on this model to analyse the effect of damping control and power system stabilisers (PSS) on all generators, except for generator7. A test system with two symmetric areas with five buses and two machines in each area along with an intermediate tie line, consisting of 11 buses, was reported in [104]. The symmetric structure makes the model exhibit almost identical frequencies in the two local modes.

The New England Test system(NETS) proposed in [44] comprises 39 buses and 10 generators. In this model, genertaor1 is equivalent of the New York system, a scaled version of the entire New york power system, to which the New England system is connected [105]. This model is widely used in the oscillation damping control literature. A simplified test system of the southern and eastern Australian system comprising of 59 buses and 14 generators in 5 areas was documented in [102]. The test system consists of 4 weakly interconnected areas and 10 local area modes. This model is primarily used for assessing the robust performance of PSS following a large disturbance. A reduced order equivalent of the interconnected NETS and New York power system was modelled as a 68 bus 5 area system for the power system analysis. There are four inter-area modes present in this system. The main challenge associated with this model is in the damping since the local and inter-area modes rely upon PSS, and three machines of this model are system equivalents.

The electromechanical dynamics of the nonlinear power system are described by modelling the system mathematically using differential/ differential algebraic equations. The dynamics of SMIB can be represented in the minimal form by the second-order differential equation, known as the swing equation. In order to model a multi-machine power system, differential algebraic equations are adopted. The swing equation can analyse rotor angle instability in SMIB and the coupled multi-machine system [106]. The rich dynamical behaviour exhibited by the swing equation makes it a candidate choice for the investigation of power system dynamics. The complex dynamical behaviour of systems analogous to SMIB, such as the forced pendulum [107] and Josephson's junction [108] are reported in the literature for different operating regimes. Therefore, as a first approximation, we adopt SMIB for our investigations, despite the simplicity of the model. The schematic representation of SMIB is shown in Fig.2.1. However, the order of the swing equation varies from two to fifteen depending upon the complexity of the problem intended to study.

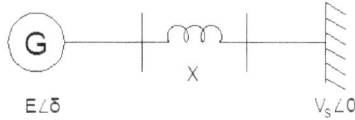

Figure 2.1: Schematic representation of single machine connected to an infinite bus. Here, the generator induced e.m.f, E leads the terminal voltage, V by an angle of δ. For generality, the load is assumed to be inductive.

Subbarao *et al.* established the onset of homoclinic bifurcations and infinite period bifurcations for variations in the damping level using a second-order swing equation representation of

SMIB power system model [49]. The presence of a stable fixed point, a limit cycle and a hysteresis region are reported in the paper for variations in damping. The effect of Gaussian white noise on the erosion of the safe basin is reported in a second-order SMIB power system model in [70]. Further, second-order SMIB is used to compute the minimum damping coefficient to ensure stability [109] . The relationship between the boundary of controlling unstable equilibrium method(BCU) and the minimum damping coefficient is established in [109]. Further, the same second-order swing equation model of SMIB is used for the computation of minimum damping in the computation of boundary of controlling unstable equilibrium method(BCU) to determine stability. Periodic and chaotic motion for the sinusoidal variation of mechanical power was observed in a second-order SMIB in [110]. Later, Ma *et al.* investigated the effect of change on the stability regimes for variations in damping [106] in a second-order SMIB model. Further, Ma *et al.* remodelled the power system by a third-order differential equation. The third-order model incorporates single-axis flux decay for a system with no damper windings. The qualitative difference in the dynamics of the second and third-order models revealed the need for varying the modelling order with respect to the research problem.

The existence of four different attractors; a stable fixed point, a stable limit cycle, and two strange attractors is illustrated in a fourth-order SMIB power system model in [111]. The results indicate that even the simplest power system model exhibits exceptionally complex dynamics. There are chances for the system to get trapped in the chaotic oscillations for a significant disturbance despite a stable equilibrium point. Abundant literature can be seen on the variations in the modelling orders of the system [112], [109]. Our investigation focuses on the non-autonomous variation of the system parameters on the stability regimes. Therefore, a third-order SMIB that retains the essential dynamics is considered for the study.

2.1.1 Mathematical model of the power system

The electro-mechanical dynamics of the rotor can be represented using the second order derivative of the rotor angle, as shown below.

$$M\ddot{\delta} + D\dot{\delta} = P_m - P_e \tag{2.1}$$

27

Here, δ is the rotor angle of the synchronous generator and ω is the rotor speed relative to the synchronously rotating frame, M represents the moment of inertia, and D is the damping factor. P_m is the mechanical power input to the system, and P_e is the electrical power output. The variations occurring to the mechanical power are neglected during the short-term stability studies [106]. However, the mechanical power varies as a function of time in physical systems [113]. In our study, the non-autonomous variation of mechanical power is modelled. The electromechanical interaction between the rotating field and the stator windings in a machine where damper windings are absent is considered to incorporate the flux decay as in Eqn.2.2

$$E_f = E_q + T_{d0}\dot{E}'$$ (2.2)

Here, E_f denotes the excitation to the field winding, E_q is the quadrature axis voltage, T_{d0} is the direct axis open circuit transient time constant, and E' is the voltage behind the transient reactance. The vector ralationship between the voltage and current along the quadrature axis can be expressed as given below.

$$E_q = E' + (x_d - x'_d)I_d$$ (2.3)

$$I_d = \frac{E' - V_s \cos \delta}{x_{d'\Sigma}}$$ (2.4)

$$I_q = \frac{V_s \sin \delta}{x_{q\Sigma}}$$ (2.5)

where I_d, and I_q denote, the currents flowing along the direct and quadrature axes of the synchronous generator, and V_s is the voltage at the infinite bus. The effective reactance along the direct and quadrature axis can be represented as

$$x_{q\Sigma} = x_q + x_T + x_L$$ (2.6)

$$x_{d'\Sigma} = x_d + x_T + x_L$$ (2.7)

The active power, P_e produced at the synchronous generator can be expressed as given below:

$$P_e = \frac{EV_s \sin \delta}{x'_d\Sigma}$$ (2.8)

28

Substituting Eqn.2.8 to Eqn.2.1,

$$M\ddot{\delta} + D\dot{\delta} = P_m - \frac{EV_s \sin \delta}{x'_d \Sigma} \qquad (2.9)$$

Further, Eqn.2.5 is substituted to Eqn.2.2

$$T_{d0}\dot{E}' = E_f - E' - \left(\frac{E'(x_d - x'_d)}{x'_d \Sigma} \right) + \left(\frac{V_s \cos \delta (x_d - x'_d)}{x'_d \Sigma} \right) \qquad (2.10)$$

The following substitutions are given for the convenience of representing the system dynamics.

$$\frac{1}{x'_d \Sigma} = B \qquad X_1 = x_d - x'_d \qquad (2.11)$$

Further, we rewrite the system dynamics in terms of first order state variables as given below:

$$\dot{\delta} = \omega \qquad (2.12)$$

$$\dot{\omega} = \frac{P_m M}{MEVB} - \frac{DM}{MEVB}\frac{d\delta}{dt} - \sin \delta \qquad (2.13)$$

$$\dot{E}' = \frac{E_f T_{d0}}{T_{d0} V_s B X_1} - \frac{E' T_{d0}}{T_{d0} V_s B X}(1 + X_1) + \cos \delta \qquad (2.14)$$

The non-dimensionalization constants in terms of the original system parameters are given below.

$$\tau = t\sqrt{\frac{EVB}{M}} \qquad \bar{D} = D\sqrt{\frac{1}{BMEV}} \qquad \bar{P}_m = \frac{P_m}{EVB} \qquad \bar{E}_f = \frac{E_f}{X_1 BV_s} \qquad \alpha = \frac{1 + X_1}{V_s B X_1} \qquad (2.15)$$

The dimensionless form of the swing equation in the following form will be used to represent the system dynamics in this study.

$$\left. \begin{array}{l} \dot{x}_1 = x_2 \\ \dot{x}_2 = -\bar{D}x_2 + \bar{P}_m - \sin x_1 \\ \dot{x}_3 = \bar{E}_f - \alpha x_3 + \cos x_1 \end{array} \right\} \qquad (2.16)$$

where,

$$X = \begin{bmatrix} x_1 \\ x_2 \\ x_3 \end{bmatrix} = \begin{bmatrix} \delta \\ \omega \\ E' \end{bmatrix} \tag{2.17}$$

In the present study, the system of equations for the deterministic system and stochastic systems are analysed. Further, the deterministic fourth-order Runge-Kutta solver is used to integrate the system dynamics in time. Then, the stability regimes of the deterministic system are identified. The stochastic modelling of the system and the numerical integration is explained in the next section.

2.1.2 Modelling of stochastic disturbances in the power system and numerical integration

This section will explain the modelling of stochastic disturbances in the power system model. The stochastic Runge-Kutta method is used to integrate the system dynamics in time. Additive white Gaussian noise(AWGN) is used to model stochastic disturbances. The intermittent feed-in from the wind and solar generations is modelled as alpha-type Levy noise in [114], [68]. However, in this work, AWGN is considered a first approximation, in line with the modelling of stochastic changes from loads and generators [82].

The increments in the white noise are generated from a Weiner process. The system of equations after incorporating the noise is given below.

$$d\delta = \omega dt \tag{2.18}$$

$$d\omega = (\bar{P}m - sin\delta - \bar{D}\omega)dt + \beta dW(t)$$

$$dE' = \bar{E}_f - E' \frac{(1+X)}{V_s BX_1} + \cos\delta + \beta dW(t)$$

where $W(t)$ is the Weiner process used to generate Gaussian noise and β is the noise intensity.

The step-by-step procedure to integrate a general stochastic differential equation using stochastic Runge-Kutta methods is explained below. Consider the general form of a stochastic autonomous differential equation for the illustration.

$$dX(t) = f(X(t))dt + g(X(t))dW(t) \tag{2.19}$$

Here, t denotes the time, which is the independent state variable, $X(t)$ is the dependent state variable, and f and g are functions of the state variable X. X can be multidimensional, depending upon the number of state variables. The noise increments, denoted by $dW(t)$, are generated from a Wiener process, which is associated with the second term of the right-hand side of the differential equation.

The dependent variable $X(t)$ at any time is evaluated using the expression,

$$X(t) = X(0) + \int_0^t f(X(s))ds + \int_0^t g(X(s))dW(s) \tag{2.20}$$

We consider a step size of δt over a total number of N time steps for the numerical integration. The value of the dependent variable τ_j at any time, where $\tau_j = j\delta t$ is [115]

$$X(\tau_j) = X(0) + \int_0^{\tau_j} f(X(s))ds + \int_0^{\tau_j} g(X(s))dW(s) \tag{2.21}$$

Similarly,

$$X(\tau_{j-1}) = X(0) + \int_0^{\tau_{j-1}} f(X(s))ds + \int_0^{\tau_{j-1}} g(X(s))dW(s) \tag{2.22}$$

Subtracting Eqn.2.22 from Eqn.2.21,

$$X(\tau_j) = X(\tau_{j-1}) + \int_{\tau_{j-1}}^{\tau_j} f(X(s))ds + \int_{\tau_{j-1}}^{\tau_j} g(X(s))dW(s) \tag{2.23}$$

In chapter 3, where we consider the deterministic system, we discard the third term on the right-hand side of the equation. The integration of the second term on the right-hand side of Eqn. 2.23 is performed using deterministic Runge-Kutta methods.

$$\int_{\tau_{j-1}}^{\tau_j} f(X(s))ds = \frac{(K_1 + 2K_2 + 2K_3 + K_4)}{6} \tag{2.24}$$

Here, the K_1, K_2, K_3, K_4 are constant increments computed using fourth order Runge-Kutta scheme. We employ principles of stochastic calculus, utilising Ito's formulation to evaluate the third term on the right-hand side of Eqn.2.23.

$$\int_0^T g(X(t))dW(t) \approx \sum_{j=0}^{N-1} \{g(X(t_j))[W(t_{j+1}) - W(t_j)] + \frac{1}{2}g(X(t_j))g'(X(t_j))[W(t_{j+1}) - W(t_j)]^2 - \delta t] + \ldots$$

$$(2.25)$$

Eqn.(2.25) modifies to Eqn. 2.26 when the coefficient of noise is independent of the state of the system.

$$\int_0^T \sigma dW(t) \approx \sum_{j=0}^{N-1} \sigma[W(t_{j+1}) - W(t_j)] \qquad (2.26)$$

If we substitute Eqn.(2.26) and Eqn.(2.24) to Eqn.(2.23)

$$X(t_j) = X(t_{j-1}) + \frac{K_1 + 2K_2 + 2K_3 + K_4}{6} + \sigma[W(t_j) - W(t_{j-1})] \qquad (2.27)$$

Now, we can numerically integrate Eqn.(2.20) to obtain $X(t)$

The state variable $X(t)$ is plotted against the system parameters to develop a bifurcation diagram. The bifurcation diagram illustrates the stability regimes and bifurcation points. The bifurcation point is the point at which a sudden qualitative change in the system dynamics is associated with a small asymptotic variation in the control parameter. A knowledge about the standard bifurcations occurring in dynamical systems is explained in the next section. A basic understanding of these bifurcations will help us identify the nature of the bifurcation that the power system model is undergoing.

2.2   Standard bifurcations in dynamical systems

Complex dynamical systems that range from expanding universe [116] to quantum systems [117] and everything in between exhibit vastly different levels of complexity. However, all these systems evolve according to a dynamical law, a specific mathematical function that governs evolution. A generic representation for a dynamical system is:

$$\frac{dX}{dt} = F(X, Y, t) \qquad (2.28)$$

where X, Y represent the vector of system state and system parameters.

The evolution of the state trajectories in the phase space provides a complete understanding of the dynamical system. Phase space of a dynamical system is a hypothetical space constructed using state variables of the system as coordinates. The trajectories in the phase space of the system evolve and eventually get attracted to an *attractor*. The asymptotic behaviour of the system can be deduced by the attractors present in the phase space of the system. Depending upon the dynamics, the attractor present in the phase space can be a fixed point, limit cycle or a strange attractor. Bifurcations can trigger the inception of new attractors or the removal of the existing attractors. Complex dynamical systems exhibit standard bifurcations, irrespective of the change in dynamics. The normal form equations provide a framework for characterising the bifurcations associated with the system.

The bifurcations observed in a power system model along with the normal forms is described in the following subsections.

2.2.1 Saddle node bifurcation

Saddle node bifurcation or fold bifurcation occurs when two equilibrium points are either created or destroyed upon the variation of the bifurcation parameter. The bifurcation results when the critical equilibrium has one zero eigenvalue. The norm of the saddle-node bifurcation is

$$\frac{dx}{dt} = \mu + x^2 \qquad (2.29)$$

where x is the state variable and μ is the control parameter. Upon varying μ in equation 2.29, we can see that the number of solution vary from no solutions to a single solution and then to two solutions as we vary μ from negative to positive. For $\mu > 0$, we have two solutions for $x_{eq} = \pm\sqrt{\mu}$. The stability of the two equilibrium solutions are determined by evaluating the

33

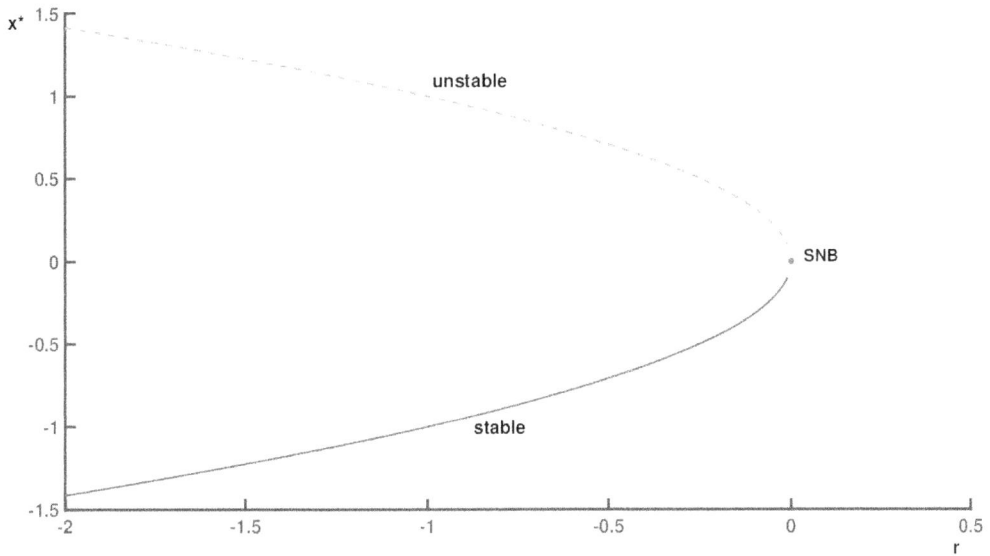

Figure 2.2: The figure depicts a saddle-node bifurcation. The stable solution is shown in the blue coloured trajectory, whereas the unstable solution is shown in the dashed red coloured trajectory.

Jacobian of the system at the equilibrium point. Among the two solutions, $x_{eq} = -\sqrt{\mu}$ will be stable and $x_{eq} = \sqrt{\mu}$ will be unstable. Fig.2.2 given below shows the saddle-node bifurcation. This bifurcation is also known as 'blue sky bifurcation ' as the two solutions appear to arise out of nowhere in phase space.

2.2.2 Hopf bifurcation

The bifurcation associated with the birth of limit cycle oscillations from a stable fixed point is termed 'Hopf bifurcation'. The minimum phasespace dimension for Hopf bifurcation to occur is two, since Hopf bifurcation is associated with oscillatory instability. The system will be in the stable fixed point when the eigenvalues are on the left half of the complex plane. Hopf bifurcation occurs when a pair of eigenvalues with non-zero real parts cross the imaginary axis.

Hopf bifurcation can be classified as supercritical or subcritical depending on the nature of the limit cycle oscillations generated. The normal form equations associated with a supercritical

Hopf bifurcation are

$$\dot{r} = \mu r - r^3$$

$$\dot{\theta} = \omega + br^3 \tag{2.30}$$

Here, r and θ represent the state variables of the system, and ω denotes the frequency of the oscillatory state.

We can observe a stable spiral(Fig.2.3(a)) for $\mu < 0$ and which is changing to a weak spiral as the μ value is increased(Fig.2.3(b)). For $\mu > 0$, the fixed point at the origin is an unstable spiral, and a limit cycle is generated at $r = \sqrt{\mu}$. In terms of flow in phase space, a supercritical Hopf bifurcation occurs when a stable spiral changes to an unstable spiral surrounded by an elliptical limit cycle [118].

The Hopf bifurcation will be subcritical if the cubic nonlinearity is destabilising. The normal form equations associated with a subcritical Hopf bifurcation are

$$\dot{r} = \mu r + r^3 - r^5$$

$$\dot{\theta} = \omega + br^3 \tag{2.31}$$

For a subcritical Hopf bifurcation, there exist two stable states; a stable spiral for $\mu < 0$ at the origin and a stable limit cycle away from the origin. Here, the stable fixed point and stable limit cycle are demarcated by an unstable limit cycle. As the control parameter is increased, the unstable limit cycle dwindles, finally engulfing the fixed point at the origin. As a result, the origin becomes unstable, leaving the limit cycle as the only stable attractor present in the system. Therefore, subcritical Hopf bifurcation will be associated with sudden large amplitude oscillations.

Two limit cycles, a stable and unstable limit cycle, coalesce and annihilate. Therefore, sub-critical Hopf bifurcation is also known as saddle-node bifurcation of cycles. Upon solving equations 2.31, we get equilibrium points of the system corresponding to limit cycle oscillations. At origin, the equilibrium point will be a stable spiral for $\mu < -1/4$. A stable fixed point and a stable limit cycle coexist between $[-\frac{1}{4}\ 0]$. In a bistable power system, we observe saddle-node

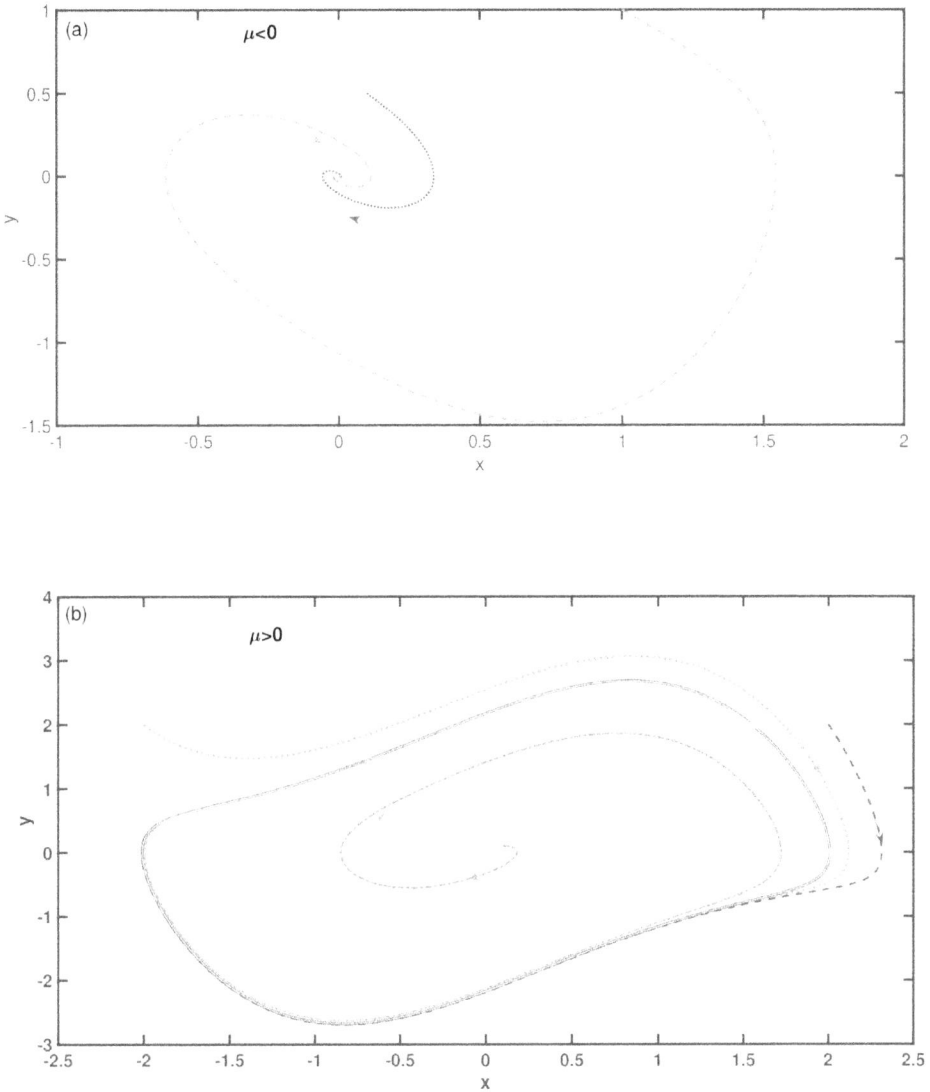

Figure 2.3: The figure plots the phase portraits for μ above and below the bifurcation point-corresponding to (2.3). The origin is a stable spiral when $\mu < 0$. For $\mu > 0$ there is an unstable spiral at the origin and a stable limit cycle, towards all the asymptotic trajectories converge.

bifurcation of cycles that results in the death or birth of unstable and stable limit cycles [119]. The Hopf bifurcation observed in the numerical experiments on the power system model is shown in Chapter 3.

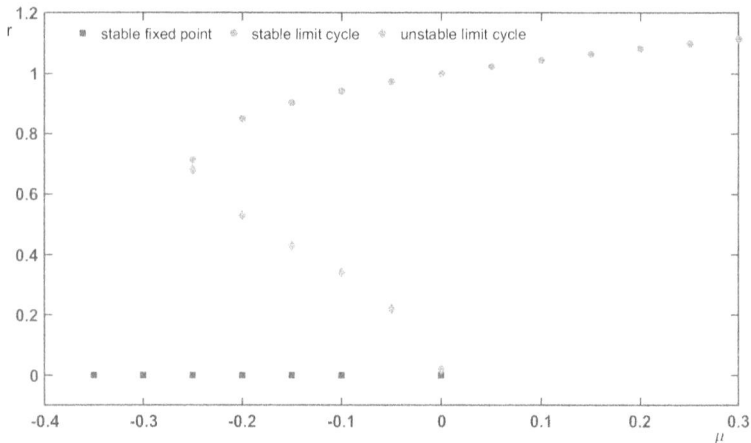

Figure 2.4: The bifurcation diagram of the system given in equation2.31 for $a = b = 1$ is depicted in the figure. The variation of r as a function of μ is shown in the bifurcation diagram. The stable fixed point of the system is indicated by the square marker and the stable limit cycle by the bullet points. The diagonal symbols represents the unstable limit cycle. The Hopf and fold point occurs at $\mu = 0$ and $\mu = -0.25$, respectively.

The oscillatory state associated with subcritical Hopf bifurcation is undesirable in power systems. The limit cycle oscillations occur in a physical system due to a pole slippage which is not advisable in the long run [49]. Hence, it is desirable to identify the stability regimes of a power system using bifurcation analysis. The following section describes the construction of a numerical bifurcation diagram of dynamical systems. The normal form and the associated dynamics were covered here near the transition point since the dynamic of complex systems simplifies down to a limited number of "normal forms", which determine the qualitative behaviour of these systems [120].

2.3   Numerical bifurcation diagram: Illustration

The development of the bifurcation diagram that demarcates the boundary of the basin of attraction of different states of dynamical systems is explained in this section. Consider, the dynamics of the power system model depicted in equation.2.16 to demonstrate the construction of the bifurcation diagram. The literature unveiled the presence of different attractors for different operating regimes of the system [111], [106] and bi-stability [49] . A bi-stable system gets its name due to the co-existence of two stable attractors. In the power system model, the attractors are a stable fixed point and a stable limit cycle, which are separated by an unstable limit cycle. Our study does not use software such as AUTO and MATCONT for the continuation analysis. Instead, we use a parameter scan to obtain the bifurcation diagram. The steps involved in the construction of the bifurcation diagram are summarised below.

1. Select an initial condition from the basin of attraction of the fixed point. This is performed by checking the asymptotic response of the state variable angular velocity to a random initial state(X_0) and for a predefined P_m value. If the asymptotic response settles to the fixed point, the initial condition is within the basin of attraction of the fixed point.

2. Perform quasi-static increments of the bifurcation parameter and capture the asymptotic response of the state variable of interest. In the experiment, the bifurcation parameter, mechanical power(P_m) is varied from 0.3 pu to 0.9pu.

3. Obtain the time series of the state variable ω corresponding to each P_m value.

4. Record the asymptotic value of ω corresponding to each P_m value.

5. Determine the point at which the ω starts exhibiting limit cycle oscillations. This point is identified as the 'Hopf point'(P_{mCT1}).

6. Increase P_m further after the Hopf point to identify the trend in the amplitude of the oscillations.

7. After procuring sufficient observations, change the initial conditions to a value belonging to the basin of attraction of the limit cycle. This is performed as explained in Step 1. However, the response is ensured to exhibit oscillations.

8. Further, decrement P_m quasi-statically from 0.9 to determine the P_m value at which the system regains the stable fixed point. This point is known as the 'fold point'(P_{mCT2}) of limit cycle.

Upon performing the procedure, a forward path(in the increasing direction of P_m) and a return path (in the decreasing direction of P_m) with respect to P_m are plotted. This is the bifurcation diagram. The Hopf and fold points are marked in the bifurcation diagram

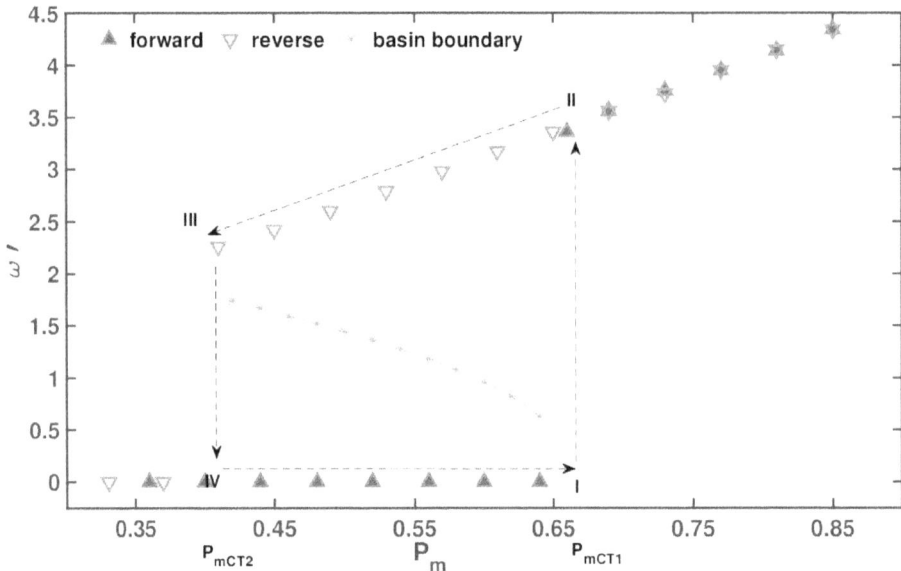

Figure 2.5: Bifurcation diagram: The bifurcation diagram depicts the RMS value of the ω for the quasi-static variation of the mechanical power, P_m. The forward transition from the stable non-oscillatory state (lower branch) to the stable oscillatory state (upper branch) occurs at the Hopf point $P_{mCT1}(\text{I})$. The reverse transition from the upper branch to lower branch occurs at the second critical point, P_{mCT2}.

at the earlier identified values. The region between the fold and Hopf point is the bi-stable region. Even though the bifurcation diagram gives information about the stability regimes, the information about the system behaviour within the bi-stable region is incomplete. In order to identify the system behaviour within the bistable region, knowledge about the basin of attraction of each state is required.

The procedure for identifying the basin of attraction of each state is outlined below.

1. Start the numerical experiment from the basin of attraction of fixed point as explained earlier.

2. In this experiment, the P_m value is varied from the fold point to the Hopf point, which is the bi-stable region. Choose the initial value of P_m as the fold point.

3. Increase the initial conditions of the state variable ω for the predefined P_m value until we get an oscillatory response for ω.

4. Record the initial condition(ω_0) that triggers ω to transit from the stable non-oscillatory state to stable limit cycle oscillations.

5. Increase the P_m value quasi-statically over the entire bistable region and repeat steps 1, 3 and 4.

6. The recorded values serve as the boundary for the basin of attraction of the stable fixed point for each P_m within the bi-stable region.

Fig.2.5 above shows the numerical bifurcation diagram that demarcates the basin of attraction of the fixed point and the limit cycle. The sustained oscillations in the power system cause severe damage to the costly equipment. Hence, developing early warning signals for the transitions in the power system is highly desirable. The following section describes the development of critical slowing down based-early warning signals from the time series.

2.4 Mechanism of early warning measures based on critical slowing down

This section provides a description of early warning measures based on the theory of critical slowing down associated with critical transitions in the time series data. Complex nonlinear dynamical systems such as ecosystems [121], biological systems [122], and financial markets [123] exhibit critical transition. The critical threshold associated with such transitions corresponds to catastrophic bifurcations in dynamical systems. Dynamical systems make irreversible transitions from a stable state to an alternate stable state while undergoing a catastrophic bifurcation. The forward and reverse transitions in these catastrophic bifurcations occur at different points resulting in a hysteresis loop [77].

A critical transition in the financial market may lead to an undesirable situation such as a market crash [123], whereas in ecological systems, it may result in species extinction [121]. The undesirable state resulting from a critical transition demands the development of early warning measures to identify the closeness of the system to the point of critical transition. However, it is difficult to predict the occurrence of critical transitions in a physical system by monitoring the state of the system as there are no significant changes before the point of critical transition. Nonetheless, recent investigations unveiled the occurrence of a comprehensive class of generic symptoms as the system approaches a critical point, regardless of the differences in the dynamics of the system. The generic symptoms can be narrowed down to the fact that sharp transitions in complex systems are related.

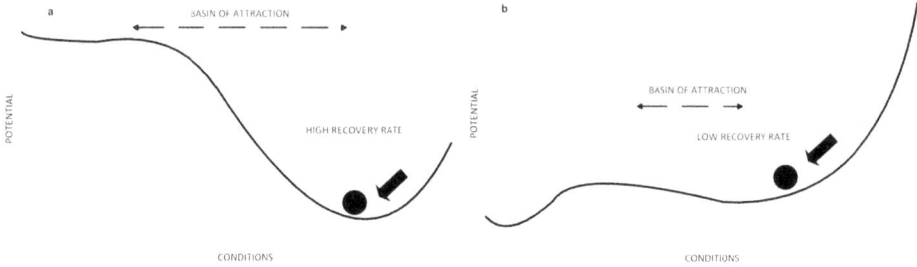

a BASIN OF ATTRACTION b

POTENTIAL

HIGH RECOVERY RATE

POTENTIAL

BASIN OF ATTRACTION

LOW RECOVERY RATE

CONDITIONS   CONDITIONS

Figure 2.6: The resilience of the system to perturbations for two different operating points (a) Basin of attarction is large and rate of recovery from perturbation is relatively high for points away from the critical point and (b) Basin of attarction is small and rate of recovery from perturbation is relatively low for points near to the critical point.

The time series for the demonstration of critical slowing down based-early warning signals are generated from the power system model described in the previous section. The eigenvalues of a dynamical system determine the stability of the system. If the eigenvalues are on the left half of the complex plane, the resulting dynamical state of the system will be stable. The dominant eigenvalue that characterises the rate of change of the state approaches zero near the Hopf point. Therefore, the states of the system become increasingly small to recover from the perturbations near the critical point. Fig. 2.6 shows the basin of attraction for the system near and away from the Hopf point. The basin of attraction shrinks in Fig.2.6(b), which will reduce the ability to recover from perturbations. As a result, there is a slowing down to recover the previous state, leading to longer memory of the states. The prolonged memory of the states is characterised by a larger standard deviation and a stronger correlation between subsequent states.

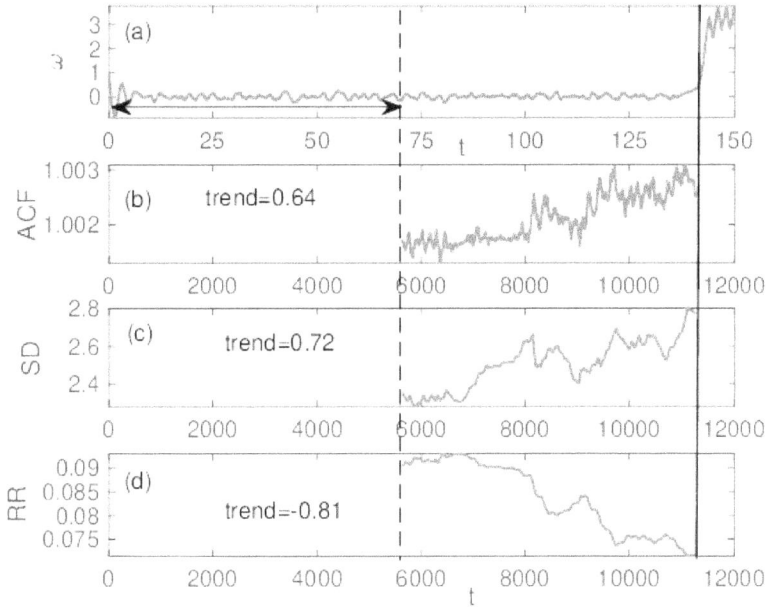

Figure 2.7: Early warning indicators of critical transition associated with subcritical Hopf bifurcation (a)Time series illustrating the shift from the non-oscillatory to the oscillatory state for the continuous variation of mechanical power from 0 to 0.8 at a rate of 0.0001(b)the autocorrelation at lag 1(c) the standard deviation(d) the return rate of the system while nearing a critical transition.

2.4.1 Development of early warning measures for subcritical Hopf bifurcation in power system

This subsection aims to familiarise the development of CSD-based early warning indicators. The early warning signals will equip the operator to take ameliorative measures to avoid imminent catastrophe. Here, the control parameter, mechanical power, is varied as a function of time, as in the case of physical systems. The mechanical power is varied at a rate of 0.0001, and the corresponding time series of angular velocity is noted down. (Fig.2.7(a)). The "Early Warning Signals Toolbox" available in R is utilised to develop the early warning signals. The trend in the statistical parameters is computed using a rolling window as in [77]. Here, 85% of the time series before the critical transition point is utilised to develop early warning indicators. A significant increase in the trend of autocorrelation and variance and a decremental trend in return rate are obtained, indicating an impending critical transition(Fig.2.7(b), (c), (d)).

2.5   Concluding remarks

This chapter describes the mathematical model used for the power system stability analysis and the rationale for choosing this model initially. The commonly occurring bifurcations in power systems, along with the normal form equations, are explained in the second section. Further, the development of a numerical bifurcation diagram is illustrated. Finally, the development of critical slowing down based-early warning instabilities to predict transitions occurring in the power system is also presented.

Chapter 3

Rate-dependent transitions in power systems

The stability studies in power systems consider the system as an autonomous system whose system parameters don't vary as a function of time. However, the power system is a non-autonomous system in which the system parameters are time-dependent. Furthermore, in renewable integrated modern power systems, the demand and generation vary with respect to time of the day. Literature evidence established that stability regimes of autonomous and non-autonomous systems are different. In this chapter, we model a non-autonomous power system model and compare the stability regimes of autonomous and non-autonomous systems. For the same, mechanical power is chosen as the bifurcation parameter. The influence of the variation of mechanical power on the quasi-static bifurcation characteristics of the power system model is presented in this chapter initially. For the same, we performed the numerical experiments by varying the bifurcation parameter while maintaining the other system parameters, such as damping and inertia constant. The variation of the time series, with respect to the bifurcation parameter, mechanical power, is plotted to obtain the bifurcation diagram. Further, the influence of the non-autonomous variation of the mechanical power on the bistable characteristics and stability regime is investigated. We also investigate the possibility of rate-induced transitions in power systems which may reduce the available operating margin and lead to instability.

3.1 Quasi-static bifurcation characteristics and stability regimes

The stability regimes of the autonomous power system model are investigated in this section with the help of a quasi-static bifurcation diagram. For the same, mechanical power is chosen as the bifurcation parameter. Then, the effect of variation of mechanical power on the bi-stable characteristics and stability regimes of the power system model is presented. We fix standard values for the system parameters given below as in the literature [111], [106]:

$$M = 0.3 \qquad\qquad B = 1 \qquad\qquad T_{d0} = 2$$

$$D = 0.2 \qquad\qquad E_f = 1 \qquad\qquad x_d/\Sigma = 1$$

The system dynamics in terms of the nondimensionalized parameters [106] are:

$$\dot{\delta} = \omega \,, \tag{3.1}$$

$$\dot{\omega} = -3.33E'sin\delta - 0.66\omega + 3.33P_m \,, \tag{3.2}$$

$$\dot{E'} = 0.5cos\delta - E' + 0.5 \,. \tag{3.3}$$

The experiment is performed by varying the mechanical power quasi-statically, both in the increasing and decreasing directions. Mechanical power input of 0.1 pu is given initially to the deterministic power system model for a predefined random initial condition. Further, Runge-Kutta fourth order scheme is used for the numerical integration for a duration of 100 time steps to obtain the asymptotic state of the system. The asymptotic values of the state variables are recorded. Further, the time series of the state variables and phase space of the system are plotted. The time series and phase space indicated the presence of a stable fixed point. Therefore, values in the range of asymptotic values of the state variables are chosen for the initial conditions of the numerical experiment. This ensures that the initial conditions are from the basin of attraction of the fixed point. Thereafter, the mechanical power was increased in a quasi-steady manner in steps of 0.05 pu. The time series of ω is recorded, and the phase space of the system is plotted corresponding to each mechanical power input. The time-series and phase space of the system indicated the presence of a stable fixed point. The power increment used in the investigation is comparable to the power increments performed in the literature. Subbarao *et al.* reported the sudden transitions from the non-oscillatory state for P_m values greater than a threshold in the second-order SMIB power system model [49]. We adopt a quasi-steady increment of mechanical power to identify the threshold value at which the transition is triggered.

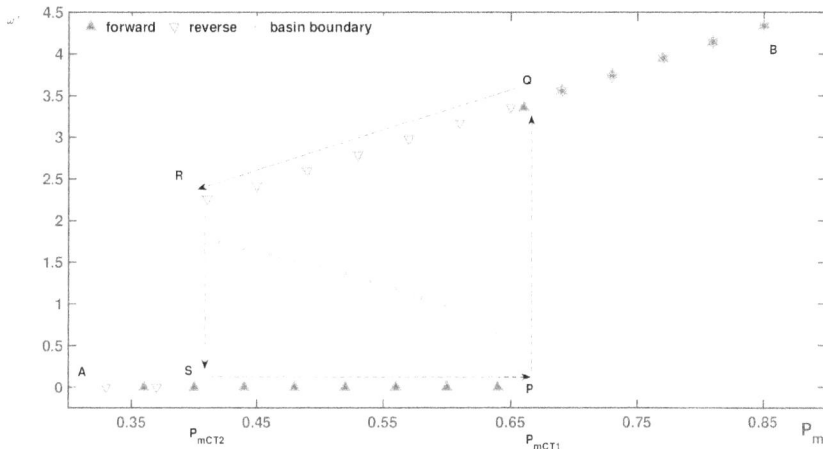

Figure 3.1: Bifurcation diagram depicting the RMS value of $\omega\prime$ for the quasi-static variation of the mechanical power, P_m. The forward transition from the stable equilibrerim point to the stable limit cycle occurs at P_{mCT1}. The reverse transition to the non-oscillatory state occurs at the fold point P_{mCT2} when P_m is decreased. The transition points are $P_{mCT1} = 0.66$ and $P_{mCT2} = 0.38$ respectively. The boundary separating the basin of attraction of stable non-oscillatory state and stable oscillatory state is demarcated by the dotted-starred line. The system can remain in the stable oscillatory or stable non-oscillatory state within the bi-stable region $P - Q - R - S$ depending upon the initial conditions. The filled triangle in blue represents the forward path, and the empty triangle in red represents the backward path. The initial conditions for the forward and reverse paths are $[\delta\ \omega\ E]$ are [0.2 0.3 0.95] and [1.2 1.5 0.9] respectively.

The root mean square (RMS) value of the angular velocity is plotted with respect to the mechanical power. It was observed that the amplitude of the RMS value of angular velocity is substantially low until the point P_{mCT1}, which corresponds to a P_m value of 0.66 pu. We continue the increment in the mechanical power to identify the increase in the amplitude of the oscillatory response. Increments of mechanical power beyond P_{mCT1} result in a sudden rise in the amplitude of the RMS value of angular velocity. The phase space of the system for P_m

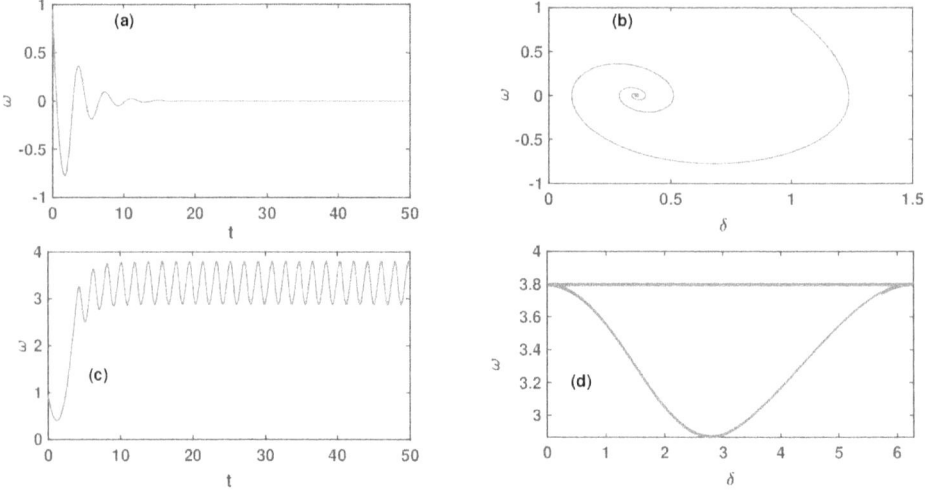

Figure 3.2: Time series and phase-space of the system in the globally stable and oscillatory region. The time series and the phase space of the system in the non-oscillatory state for mechanical input $P_m = 0.34$ is shown in Fig. 3.2(a) and Fig. 3.2(b). The time series and phase-space of the system for a mechanical power input of $P_m = 0.74$ is shown in Fig. 3.2(c) and Fig. 3.2(d). The initial conditions of the system are given by $\delta = 0.2, \omega = 0.95$, and $E = 1$.

values beyond P_{mCT1} exhibited an isolated closed trajectory marking the presence of a limit cycle. Fig.3.1 shows the bifurcation diagram obtained by varying the mechanical power. We confine the increase of mechanical power up to $P_m = 0.9$, which correspond to the point B in the bifurcation diagram. The RMS value of the state possessed by the system during the mechanical power increments is termed 'forward path'.

The initial conditions of the system are changed once the system reaches point B in the bifurcation diagram. The new initial conditions are the asymptotic values of the state variables that correspond to any of the P_m values beyond P_{mCT1}. The mechanical power is decreased in steps of 0.05 p.u. The RMS values of the states attained during the decrement of mechanical power are termed as 'return path'. The system occupies the state of stable limit cycle oscillations up to the point R in the bifurcation diagram during the return path. The system reverts to the stable non-oscillatory state at $P_m = P_{mCT2}$. The P_m value that correspond to P_{mCT2} is 0.38. The system is globally stable in the region 'AS' in the bifurcation diagram. The window from P_{mCT2} to P_{mCT1} is identified as the bistable region. The region beyond P_{mCT1} is globally unstable.

47

The presence of unique forward and return paths('ASPQB' and 'BQRSA') establishes a hysteresis region in the system. The bifurcation diagram shows the presence of two stable states in the region 'SP'; the stable fixed point and stable limit cycle oscillation being the two states. Depending upon the initial conditions, the system can be driven to the state of limit cycle oscillations, even during the forward path [118]. The system is said to be bi-stable in the region 'SP'. Hence, it is evident that the power system model analysed in this thesis exhibit both hysteresis and bi-stability.

The time series of angular velocity and the corresponding phase space for P_m value in the globally stable region and unstable region is shown in Fig. 3.2. The presence of a bi-stable region (Fig.3.1) and the limit cycle (Fig.3.2(d)) confirms the presence of subcritical Hopf bifurcation in the power system model. In order to obtain information regarding the system behaviour within the bi-stable region, the basin of attraction of each attractor is to be identified. For the same, mechanical power is varied within the bi-stable region quasi-statically. The experiment is started from initial conditions that ensure the asymptotic decay of the time series. Gradually increment the initial conditions to identify the initial condition that triggers the transition to the oscillatory state. Record the initial condition that triggers the transition for the particular mechanical power. Repeat the experiment for the entire P_m values within the bi-stable region. The recorded values serve as the boundary of the unstable limit cycle, indicated by the curve represented by the star symbol in the bifurcation diagram. The step-by-step procedure for the computation of the boundary of the basin of attraction is elaborated in Chapter2.

We present a semi-analytic solution demonstrating the transitions in the power system model.

3.2 Semi-analytic determination of the stability regimes of the power system model

The electro-mechanical dynamics of the system described in Eqn.2.16 are considered for developing the semi-analytical solution.

According to Eqn.(3.1–3.3), we get the equilibrium points upon solving the algebraic equations as in (3.4)-(3.6):

$$\omega = 0 \,, \tag{3.4}$$

$$-3.33E' sin\delta - 0.66\omega + 3.33P_m = 0 \,, \tag{3.5}$$

$$0.5cos\delta - E' + 0.5 = 0 \,. \tag{3.6}$$

The complexity of the equations (3.4)–(3.6), makes finding out an explicit expression for δ_0 and E_0 difficult. Simplifying Eqn.(3.5)–(3.6), the equilibrium points can be obtained in terms of P_m as:

$$P_m = \frac{1}{2}[\frac{1}{2}sin(2\delta) + sin(\delta)] \,, \tag{3.7}$$

$$E' = \frac{1}{2}(1 + cos(\delta)) \,. \tag{3.8}$$

The Jacobian J of the system is determined in order to analyze the stability of the equilibrium points as follows:

$$J = \begin{bmatrix} 0 & 1 & 0 \\ \\ -3.33E'_0 cos\delta_0 & -0.66 & -3.33 sin\delta_0 \\ \\ -0.5 sin\delta_0 & 0 & -1 \end{bmatrix} \tag{3.9}$$

The trace, τ and determinant, Δ of the Jacobian matrix can be obtained as:

$$\tau = -1.66 \,, \qquad\qquad \Delta = 3.33[\frac{1}{2} sin\delta_0^2 - E'_0 cos\delta_0] \,.$$

The value of τ remains constant as the damping is not varied. Therefore, the stability of the equilibrium points of the system is determined by P_m. When we vary P_m, Δ is positive to begin with. The stability of the system is analyzed by evaluating the sign of $\tau^2 - 4\Delta$ when Δ is positive. Substituting for Δ in terms of P_m for analyzing the nature of the fixed points, we

49

get the following expression:

$$\Delta = 3.33 \left[\frac{P_m^2}{2E'^2} - \left(\frac{1 - \frac{P_m^2}{E^2}' - \sqrt{1 - \frac{P_m^2}{E'^2}}}{2} \right) \right].$$

(3.10)

The system is having fixed points when Δ is real and no fixed point when Δ is imaginary [118]. Thus,

$$sin\delta_0 = \frac{P_m}{E'}, \qquad E_0' = \frac{1 + \sqrt{1 - \frac{P_m^2}{E'^2}}}{2}.$$

(3.11)

The determinant becomes an imaginary number when $\frac{P_m}{E'} > 1$. As we increase P_m, a point is reached at which P_m just crosses E', where Δ becomes imaginary, leading to vanishing of the fixed points [118]. Fig. 3.3(c) shows the point at which no fixed point exist. The P_m value at which the fixed points disappear is called the point of CT in the forward path denoted as P_{mCT1}.

We solve the characteristic equation of the system to obtain the nature of the equilibrium point. Upon solving the characteristic equation ,

$$\lambda^3 \; + \; 1.66\lambda^2 \; + \; \lambda(0.66 \; + \; 3.33E_0'cos\delta_0) \; + \; 3.33[E_0'cos\delta_0 \; - \; 0.5sin^2\delta_0] \;\; = \;\; 0 \quad (3.12)$$

we get three eigen values, of which one is a pair of complex conjugate number with negative real part and the other is a real number with negative real part. The nature of the eigen values indicates that the fixed point is a stable focus node(FN).

We adopted a strategy similar to the analysis of a forced oscillator in order to inspect the system behavior upon crossing P_{mCT1} [118]. We consider the nullcline of ω presented in Fig. 3.3(a) for $P_m > P_{mCT1}$. It is established that nullcline is the set of points to where all the trajectories of ω asymptotically converge [118]. The nullcline is presented for $P_m = 0.9$ such that P_m is greater than P_{mF} [118].

$$\omega = \frac{1}{2}[P_m - E'sin\delta].$$

(3.13)

The long term behavior of the system can be analyzed by evaluating the system in the

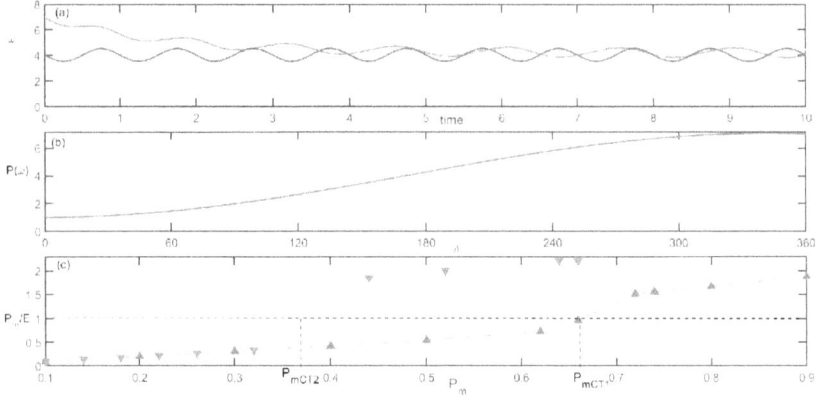

Figure 3.3: (a) Trajectories of ω converging to nullcline for $P_m > P_{mCT1}$: the curve in lavender color indicates the nullcline of the system depicted by Eqn.(3.13); the trajectory in red color shows the convergence of a system trajectory from a high initial condition to the nullcline; ω starting from a low value converging to the nullcline is demonstrated by the green trajectory. (b) Poincare map traversing from $\delta = 0$ to $\delta = 2\pi$ within the rectangular box. (c) Variation of P_m/E for P_m illustrating the flux decay for increments in P_m for initial conditions within the basin of attraction of the fixed point is shown in curve in blue color. The response for initial conditions that belong to the basin of attraction of the limit cycle is shown in coral color. The value at which these curves crosses the horizontal line of magnitude 1 is the points of CT in the forward and reverse path respectively.

phase space. We confine our investigation within a rectangular box where δ ranges from $[0\ 2\pi]$, and ω varies from a lower bound below the nullcline to an upper bound above the nullcline as shown in Fig. 3.3(b), since the limits are sufficient to analyze the asymptotic behavior. Consider $P(\omega)$, which is a trajectory in cylindrical coordinates. The trajectory illustrates us how the height of a trajectory $P(\omega)$ changes after one lap around the cylinder (δ, ω). Fig. 3.3(b) depicts the trajectory $P(\omega)$ traversing from ω_1 to ω_2. If $P(\omega)$ is such that $P(\omega) = \omega$, then it indicates the existence of periodic orbit [118]. Fig. 3.3(a) shows that trajectories of ω above the nullcline travel downwards to the nullcline, whereas trajectories of ω below the nullcline traverse upwards. Therefore, the considered $P(\omega)$ is a monotonic function. This reveals the long term behavior of the system when $P_m > P_{mCT1}$ will be a CT from stable Focus Node to limit cycle oscillations. As P_m is decreased from values above P_{mCT1} asymptotically, Δ changes from imaginary to real upon crossing a threshold value $P_m = P_{mCT2}$ (denoted by IV in Fig. 3.3). As we decrease P_m, the amplitude of the limit cycle oscillations get reduced, leading to vanishing of limit cycle oscillations at P_{mCT2}, which is known as the fold point.

The sub-critical Hopf bifurcation occurring in the power system is highly undesirable. The following section presents the influence of non-autonomous variation of mechanical power on the stability regime.

3.3 Effects of rate-dependent variation of mechanical power on the stability of power systems

Mechanical power is varied as a function of time in order to analyse the transitions in non-autonomous power systems. The perpetual variations occurring in renewable energy integrated modern power systems are modelled in the non-autonomous power systems. We model small increments to the mechanical power as a function of time in the non-autonomous power system model [124], [125]. The increments in mechanical power are in line with the literature [113]. Our model, which considers small linear increments, rules out the possibility of severe transient disturbances, which is beyond the scope of this study.

The non-autonomous variation of mechanical power is performed as in the equation given below.

$$P_m(t) = P_{m0} + \mu\, t \qquad (3.14)$$

where P_{m0} is the initial value of the mechanical power, and μ is the rate at which the mechanical power is being varied.

Fig.3.4 depicts the transitions occurring in the power system model for the non-autonomous variation of mechanical power. Fig.3.4 illustrates that the transitions from the stable non-oscillatory state to the oscillatory state occur at different P_m values for different rates of variation of mechanical power. Additionally, the delay observed in the point of transition with respect to the Hopf point is significant for faster rates. The delay in the point of transition can be attributed to the time taken to negate the accumulated effects of negative eigenvalues with that of the integrated effects of positive eigenvalues.

The delay in transition to the alternate stable state for the slow variation of the control parameter is reported in the literature. Our result on the delay in transition is in line with the literature. Further, Wieczorek et al. noticed the occurrence of rate-induced transitions for rates above a critical threshold in climate systems [59]. [59] developed the necessary and sufficient

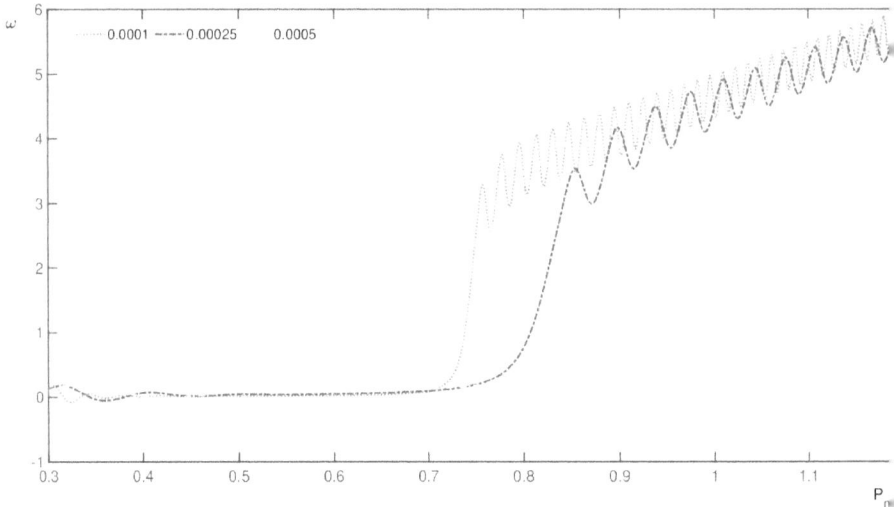

Figure 3.4: Transitions in angular velocity for the rate-dependent variation of mechanical power at three different rates. The rates considered here are $\mu_1 = 0.0001\text{pu/time step}$ $\mu_2 = 0.00025\text{pu/time step}$ $\mu_3 = 0.0005\text{pu/time step}$

conditions for the existence of critical ramping in a general class of slow-fast systems. The authors elucidated an explosive release of soil carbon from peat lands to the atmosphere for the critical rate of global warming leading to potential climate tipping termed ' compost bomb instability' with this framework. The authors also demonstrated that systems which undergo ramping variation form singularly perturbed systems with a minimum of two slow variables. Further, the authors uncovered possible phase portraits of the ramped system by exploiting terms of folded singularities and associated canard trajectories. A consecutive study by the same group of authors identified the possibility of bifurcation-induced transitions turning to rate-induced transitions for faster rate variation of a system parameter [56]. These studies revealed that the system follows the quasi-static attractor that follows a slow variation for rates below the critical rate. Tony *et al.* assimilated the possibility of rate-induced transitions in thermoacoustic systems [60]. The authors established the relationship between the critical rate and initial conditions to trigger a transition to the alternate state in thermoacoustic systems. The thermoacoustic system chosen for the study was found to undergo sub-critical Hopf bifurcation. The power system model considered also undergoes subcritical Hopf bifurcation, and the resulting oscillations are highly undesirable. Therefore, we will investigate the possibility of rate-induced transition and its effects on the stability of the power system model in the next section.

3.4   Rate-induced transitions in power systems

The rate-dependent variation of mechanical power as performed in the previous section is carried out, with a restriction for the variation of mechanical power in this section. The variation of mechanical power is limited within the initial and final values, specified as P_{m0} and P_{m1} to ensure noncrossing of the Hopf point during the study.

$$P'_m(t) = P_{m0} + \mu t \tag{3.15}$$

Here, $P'_m(t)$ represents the non-autonomous variation of mechanical power, which starts at P_{m0} and varies at a rate of μ. The time series of the state variable, angular velocity and the phase space of the system for the initial value P_{m0}(Fig.3.5(a), (b)) and the final value P_{m1}(Fig.3.5(c), (d)) between which the variation of P_m is performed are shown in Fig. 3.5.

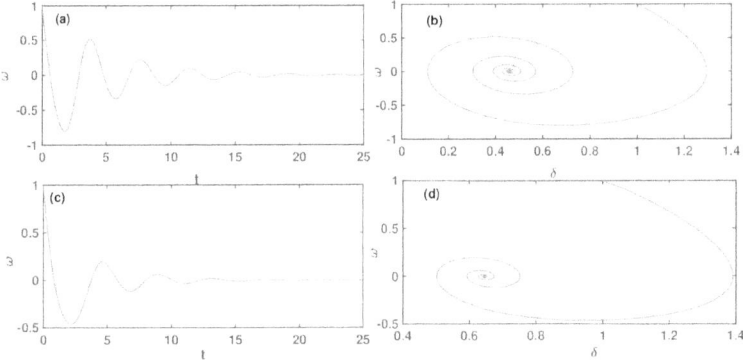

Figure 3.5: Time series and phase space of the system: Figure depicts the time series and phase space of the system for P_{m0} and P_{m1}, within which the mechanical power is allowed to vary. The time series and phase space of the system for the initial value $P_{m0} = 0.42$ is shown in Fig.3.5(a), (b). Fig. 3.5(c), (d) shows the time series and phase space for $P_m = 0.54$ that correspond to P_{m1}. The initial conditions for procuring the time series are the same as that of the forward path.

We maintained P_{m0} at 0.42 for a duration of 5 time steps. This ensures the asymptotic decay

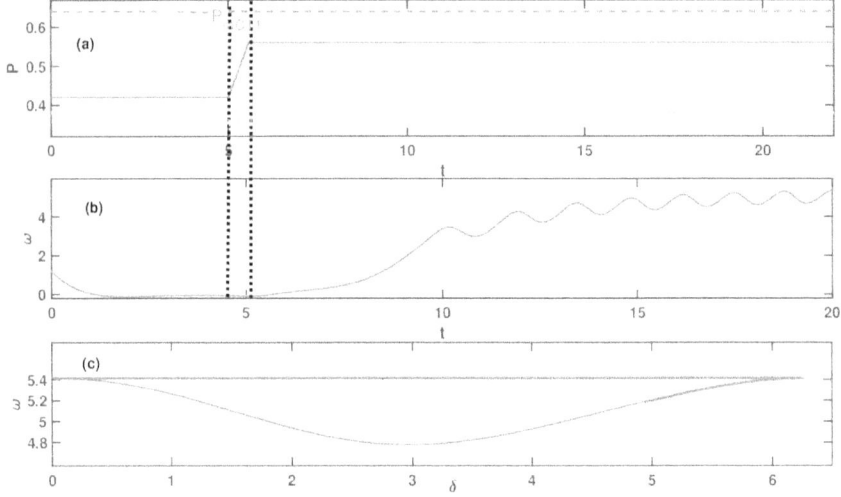

Figure 3.6: Rate induced transitions in power systems. The mechanical power varying at a rate of $\mu = 0.001$ is shown in Fig.3.6(a). The P_m values at which the Hopf and fold bifurcation occur are shown by dashed lines in Fig.3.6(a). The time series and phase space of the system are shown in Fig.3.6(b), (c)

of the states to the stable attractor. This also eliminates the possibility of a transition to the alternate state due to the collective effects of ramp variation and transient dynamics. Fig.3.6(a) shows the variation of input mechanical power as a function of time. The corresponding time series and phase space of the system are shown in Fig.3.6(b), (c). We observed that the system is driven to large amplitude limit cycle oscillations before the Hopf point during the rate-dependent variation of mechanical power. We also noticed that the rate-induced transitions are absent in the system for rates below $\mu = 0.001$ for the initial conditions chosen for the experiment.

Further, we examined the rate-induced transitions in Fig. 3.6 with respect to the quasi-static bifurcation diagram to identify the reason for the early onset of transition to the oscillatory state. Here, we examined the response of the state variable ω for the variation of P_m at the same rate as in Fig. 3.6 for two different initial conditions. The initial conditions chosen are $\omega = 1.65$ and $\omega = 0.8$. The response of the system seems to decay initially until the curve in yellow colour that starts from $\omega = 1.65$ crossover the unstable limit cycle and grows towards the stable limit cycle while reaching the final value P_{m1} (Fig.3.7). The trajectory in pink colour from the initial value $\omega = 0.8$ stays within the basin of attraction of the stable fixed point.

55

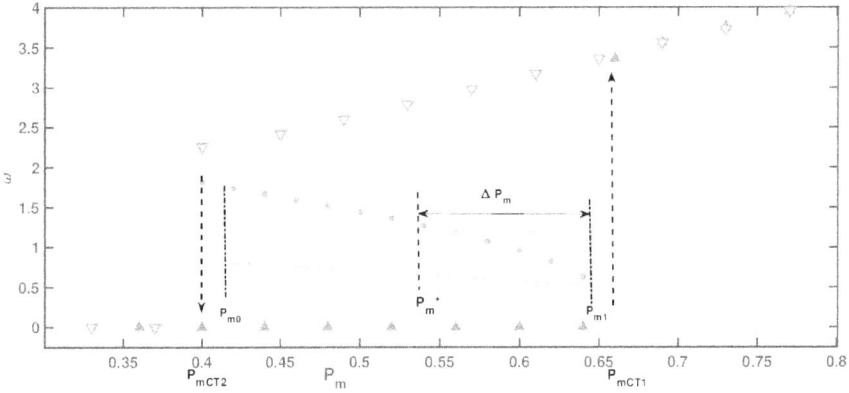

Figure 3.7: Rate induced transitions against quasi-static bifurcation. Figure depicts the RMS value of the state variable ω for the rate dependent variation of mechanical power from 0.42 to 0.54 for two different initial conditions. The curve starting from the initial condition $\omega = 1.65$, represented in yellow color crosses the unstable limit cycle, and grows towards the stable limit cycle. The point at which trajectory crosses the unstable limit cycle is marked as P_m*. The reduction in stability margin due to the early transition is denoted as ΔP_m. The second curve, starting from $\omega = 0.8$ depicts the asymptotic decay to the fixed point for the variation of mechanical power at a rate of $\mu = 0.001$.

We infer that the reason for the rate-induced transitions could be the slower decay rate of the system to the basin of attraction of the fixed point compared to the rate of variation of P_m. Moreover, the decay rate is dependent on the initial conditions. The slower decay rate of the system to the basin of attraction of fixed point with respect to the evolution of P_m is identified as the reason for the rate-induced transitions, in line with the observations by Fruchard in 'discrete ducks' [126]. Hence, we concluded that the critical rate required for the transition from the stable non-oscillatory state to the unstable oscillatory state is a function of initial conditions.

The threshold rate that triggers transitions in the system is termed as critical rate, μ_c. In order to determine the critical rate, the mechanical power is varied at different rates for a prefixed initial condition. Further, the system is allowed to evolve towards the asymptotic state. The rate at which the system transits towards the limit cycle oscillations for the predefined fixed initial conditions is noted down as μ_c for the initial conditions. The experiment is repeated for ω values in the window [1.97 2.6]. The ω values considered here is capable to transition to the limit cycle state, depending upon the values of the other state variables(δ, E'). The values of ω less than the lower limit of the interval ensure global stability. Once the upper limit of the ω is reached, irrespective of the values of the other state variables, the system response transits to the oscillatory state. Hence the area of interest is restricted to the interval. It was observed that for higher initial conditions for minimum rate for triggering a transition is lower compared to low initial conditions. The variation of critical rate with respect to initial conditions is depicted in Fig. 3.8

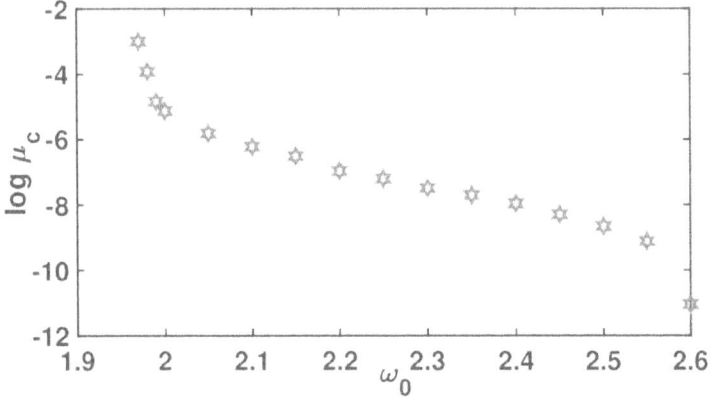

Figure 3.8: Dependence of critical rate on the initial conditions: Figure shows the critical rate μ_c that triggers rate-induced transitions in the power system model. The mechanical power is allowed to vary within the bi-stable region at different prefixed rates. The angular velocity values considered for the variation of initial conditions lie within the range [1.97 2.6]

Our study on identifying the critical rate pertaining to an initial condition will provide knowledge about the safe limit for the variation of the system parameter. This will help to ensure the operation within a stable regime. The presence of critical rate and its dependence on initial conditions observed in our power system model is in line with the observations by Tony *et al.* in thermoacoustic systems [60].

3.5   Concluding remarks

The mechanical power is varied as a function of time in the present study. The bi-stable region, presence of limit cycle oscillations, and sudden increase in the RMS value of the response at the point of transition mark the existence of a sub-critical Hopf bifurcation. Further, we identified the possibility of rate-induced transitions in a canonical power system model for rates above a critical threshold. We inferred that the slower decay rate of the system to the basin of attraction of the fixed point with respect to the rate of variation of P_m triggers rate-induced transitions. We have also established the dependence of the critical rate that triggers a transition to the alternate state upon the initial conditions.

Chapter 4

Effect of stochastic fluctuations on the stability of power systems

The paradigm shift in power systems associated with integrating intermittent renewable resources and electric vehicles has increased uncertainties in power systems. There have been innumerable efforts to analyse the stability of power systems in the presence of stochastic disturbances. The stochastic stability studies in power systems incorporate noise models such as Gaussian white noise [82], coloured noise [127], and Levy-type noise [114]. However, the effect of stochastic fluctuations in the dynamic evolution of non-autonomous power systems is unexplored. In this chapter, we numerically explore the influence of additive white Gaussian noise on the transition characteristics in the power system model. We consider both autonomous and non-autonomous power system models for our investigation to draw a comprehensive understanding of the influence of noise on the stability regimes. For the same, bifurcation experiments are performed to identify the influence of noise on the width of the bi-stable region with respect to different noise intensities. Further, we study the impact of noise on rate-dependent transitions in the power system model. Finally, we present an emergency control strategy to restrict the operation within the safe operating limits.

4.1 Quasi-static bifurcation characteristics in the presence of noise

The bifurcation diagram that illustrates the variation of RMS value of angular velocity for variation in mechanical power in the absence of noise is shown in Fig.4.1. The bifurcation diagram exhibits a distinct bi-stable region and a sudden jump in the value of RMS value of angular velocity. Further, the time series depicting sustained oscillations(Fig.4.1(b)) and phase space(Fig.4.1(c)) with an isolated closed trajectory indicating a limit cycle is presented. The transition to instability occurs via a subcritical Hopf bifurcatrion at a P_m value of 0.66 pu. [(3.1)-(3.3)]

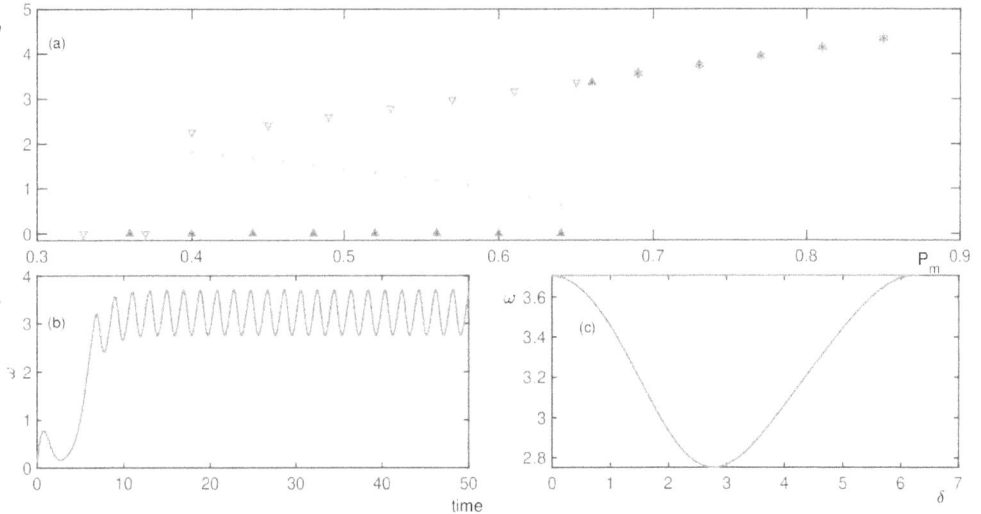

Figure 4.1: (a)The bifurcation diagram illustrates the RMS value of angular velocity for variations in mechanical power. The transition to instability occurs via a subcritical Hopf bifurcation at $P_m = 0.66$. The initial conditions for the forward and reverse paths are $[\delta \omega E]$ are [0.2 0.3 0.95] and [1.2 1.5 0.9] respectively. (b) The time-series of angular velocity that demonstrates sustained oscillations (c) Phase portrait depicting an isolated closed trajectory in the phase space. The filled blue triangle represents the forward path, whereas inverted empty triangle denotes the reverse path.

Wei *et al.* [70] reported basin erosion for noise intensity above a threshold in the literature on power systems. The authors also observed that a further increase in the noise intensity above the threshold might induce transitions to instability. Therefore, as a first step in identifying the influence of noise on the transition characteristics, we have to identify the safe upper limit on the noise so that system does not undergo noise-induced transitions. To determine the same, we investigate the effect of noise in inducing transitions by developing the quasi-static bifurcation diagram in the presence of noise. Since all the system dynamics are non-dimensionalised, the noise also must be non-dimensionalised. The non-dimensionalisation of noise is presented in the following subsection.

4.1.1 Non-dimensionalisation of noise

In order to non-dimensionalise the noise, we have given a β value of 0.03 to the system of equations given by Eqn.2.19, for a predefined mechanical power($P_m = 0.7$), and recorded the time series of angular velocity at this noise level. Further, the time series of the corresponding deterministic system is also recorded. Initially, the RMS value of angular velocity in the non-oscillatory state is measured. Let this be denoted by the intensity, I. Further, the RMS value of the limit cycle oscillations at the Hopf point in the absence of external noise is measured, denoted by I_H.

The non-dimensionalized noise intensity,

$$\sigma = \frac{I}{I_H} \tag{4.1}$$

In short, the ratio of the RMS value of limit cycle oscillations in the presence of noise to the RMS value of limit cycle oscillations in the absence of noise is the non-dimensionalised noise intensity. The non-dimensional noise intensity is denoted by the letter σ here onwards. Further, a calibration curve that illustrates the non-dimensional noise intensity, σ for the corresponding β values, is presented. The figure below shows the calibration curve.

Further, we conducted experiments in the presence of noise to investigate the influence of noise on the stability characteristics. To perform the same, we varied σ between 2% to 8%. Fig.4.2(a) shows the bifurcation diagram in the presence of noise intensity of 2%. The corresponding time series and phase space are shown in Fig.4.2 (b) and (c), respectively.

We observed that the subcritical nature of the bifurcation is not destroyed in the presence of external noise of intensity 2%. It can be seen that the forward transition occurs at an earlier point, at a P_m value of 0.64 in the presence of noise. The width of the bi-stable region decreases

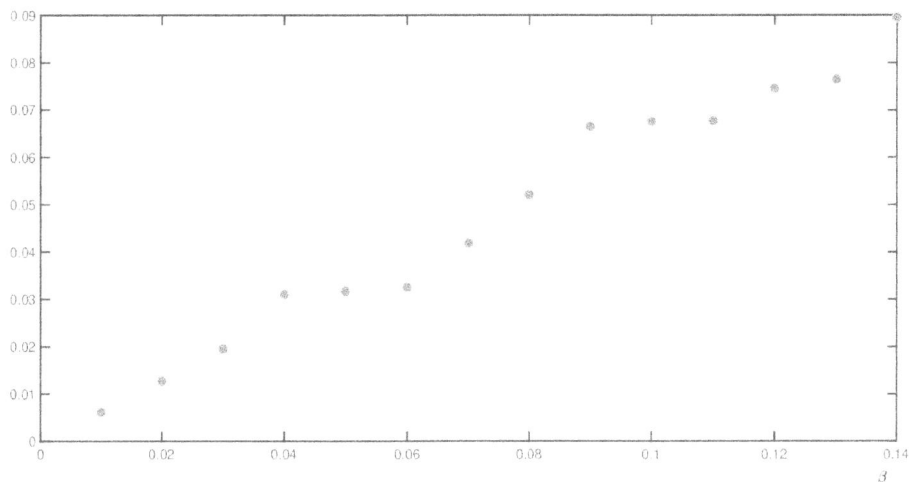

Figure 4.2: Non-dimensionalisation of noise: The calibration curve demonstrates the variation of non-dimensional noise intensity σ to the corresponding noise intensity β.

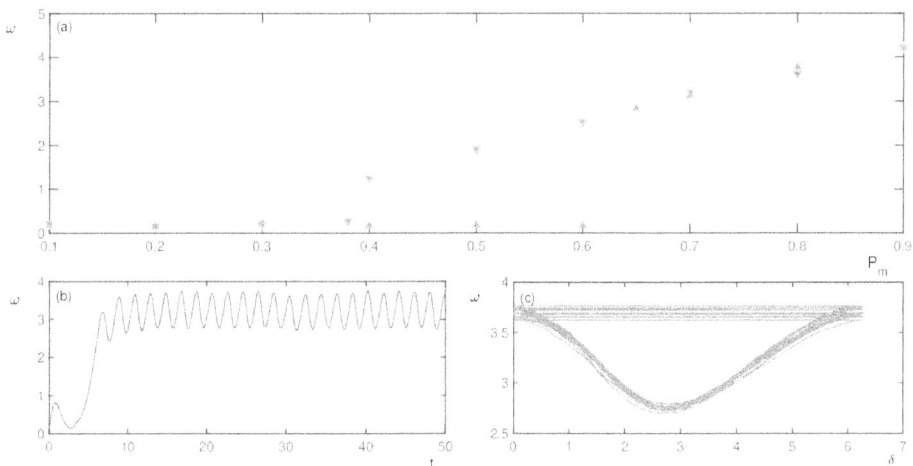

Figure 4.3: Quasi-static bifurcation diagram in the presence of noise: The figure depicts the quasi-static bifurcation diagram in the presence of external noise intensity of 2%. Here, the transition to instability occurs through a sub-critical Hopf bifurcation at a P_m value of 0.64. The forward and reverse paths are represented by filled and empty triangles in Fig.4.3(a). (b)The corresponding time series of angular velocity was captured at a P_m value of 0.64, which indicates the presence of a limit cycle at an earlier P_m value (c). The phase pace at this P_m value is a closed trajectory, which is thickened compared to Fig.4.1(c). The initial conditions are the same as in Fig. 4.1.

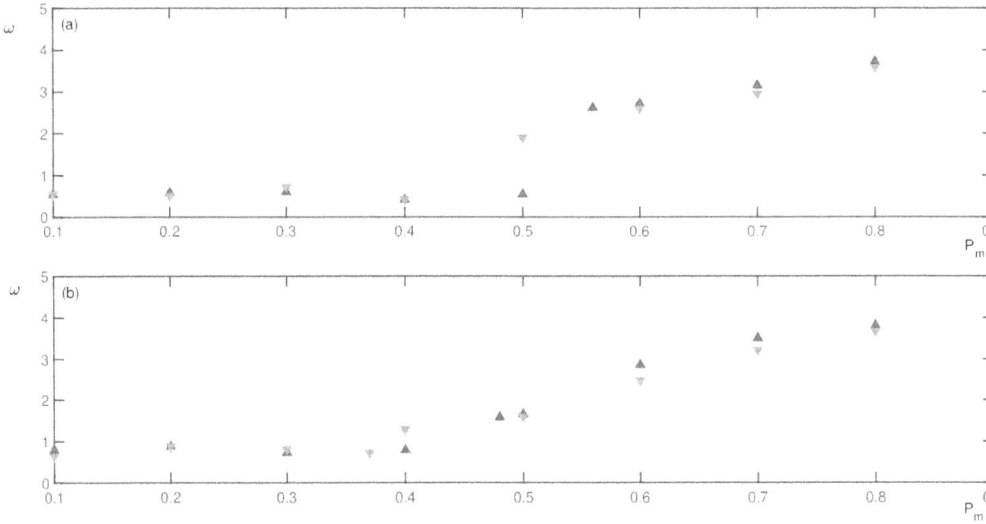

Figure 4.4: Influence of noise on the bifurcation diagram: Fig.4.4(a), (b) shows the bifurcation diagram at noise intensities of 4 and 6%, respectively. The reduction in the width of the bi-stable region for increments in noise intensities can be seen in the figure.

due to the shifting of the Hopf point. Further, the trajectories in the phase space are spread out, indicated by the increased thickness of the closed trajectory in the presence of noise. The system becoming practically unstable farther from the Hopf point of the deterministic system for increments in the intensity of noise is reported in the literature of other dynamical systems [128]. Waugh and Juniper observed the system dislodging from stable periodic solution to zero solution for an increase in the noise strength. Their work also established the decrease in the width of the bi-stable region, though an explicit statement or quantification regarding the same is absent. The reduction in the width of the bi-stable region in the presence of noise observed in the present work is nowhere reported in the literature on power systems. Nonetheless, the influence of noise on the bi-stable characteristics is reported in the context of other dynamical systems [129], [130], [131], [132]. The results indicate that the width of the bi-stable region decreases with an increase in noise intensity. However, the transition to instability was found sub-critical for all noise intensities.

Further, we observed that the transition to instability turns out to be less abrupt for noise intensities higher than 6%. We also observed that the bistable region is non longer discernible

63

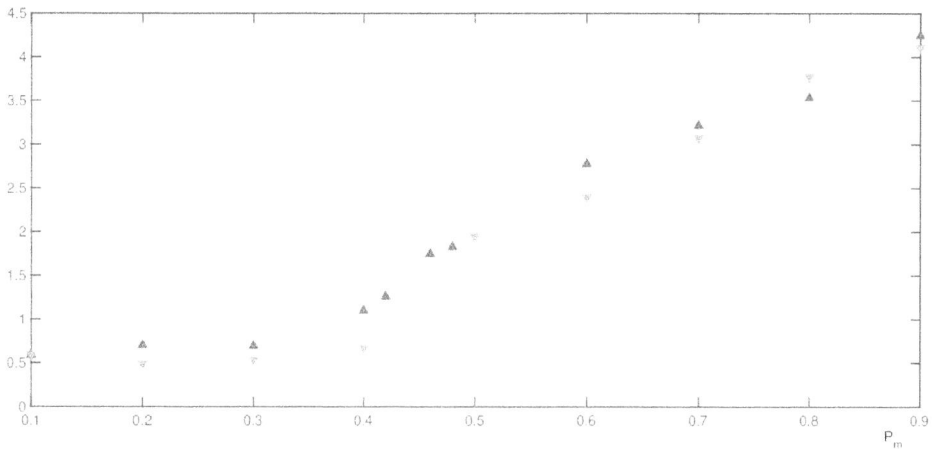

Figure 4.5: Bifurcation diagram illustrating the RMS value of angular velocity for variation in mechanical power at a non-dimensional noise intensity of 8%. We can see suppression of the bi-stable region for increments in the noise intensity.

for higher noise intensities($> 6\%$). To be precise, the transition becomes continuous for higher noise intensities. Furthermore, the exact overlap between the forward and reverse paths seemed impossible even when the transition was continuous. We reasoned that each realisation of a stochastic process is different even though other system parameters are constant.

The system oscillates between the two alternate states while operating at high-intensity noise. Scheffer *et al.* termed this phenomenon as flickering in the context of critical transition observed in ecosystems [77]. Scheffer *et al.* observed that the transition turns out to be continuous, wiping off the bistable region in the ecosystem due to flickering. The presence of a bi-stable region and sudden transition characterize a sub-critical transition. Since these two features are absent for noise intensities beyond the threshold, the transition can no longer be identified as sub-critical. Our results on the suppression of the bi-stable region are in line with the results of the non-linear oscillator models [130]. The continuous transition and loss of the bi-stable region observed at $\sigma = 8\%$ restrict the upper noise level considered for the study below this threshold, to eliminate the possibility of noise-induced triggering. The suppression of the bi-stable region for higher noise intensity (8%) is in qualitative agreement with the observation of the destabilisation of the attractors due to additive noise for the slow sinusoidal parameter variation in the logistic map [133], and annihilation of one of the coexisting attractors in a bistable system [134]. In the following section, we will investigate the effect of noise on the stability regimes of the non-autonomous power system model for noise intensities below the threshold value(8%)

4.2 Transitions in the non-autonomous power system model in the presence of stochastic fluctuations

In this section, we evaluate the response of the system for variation of system parameters as a function of time in the stochastic bi-stable power system oscillator. Even though the stability of a bi-stable power system oscillator is well investigated, these studies focus on the static bifurcation analysis of the bi-stable oscillator [106], [47]. Perpetual changes in load and generation associated with RES make modern power systems non-autonomous [135]. In our earlier work, we have established that the system continues to hover around the stable equilibrium point for a slow variation of the system parameter. Hence, the onset of oscillations is delayed with respect to the quasi-static bifurcation diagram [136]. To be specific, the destabilisation of the slowly varying trajectory does not occur immediately upon crossing the bifurcation point, determined by a quasi-static bifurcation analysis [54].

Here, we inspect the influence of stochasticity in the bi-stable power system model, where the system parameter varies as a function of time. The mechanical power varies as a function of time, as in the previous study as follows:

$$P_m(t) = P_{m0} + \mu t \tag{4.2}$$

where P_{m0} is the initial value of the mechanical power and μ is the rate at which mechanical power is varied. We allow a linear evolution of mechanical power as discussed in the literature [113]. We have paid sufficient attention to eliminate the possibility of severe transient disturbances by limiting the maximum rate at which the bifurcation parameter evolves, which is beyond the scope of this work. In our experiments, we have non-dimensionalized noise before performing the experiment as in the literature [132].

We capture the time series of the state variable ω for the investigation, which contains sufficient information about the state of the power system [137], [138]. In order to investigate the response of the system to variations in initial conditions, we consider two sets of initial conditions, one at a finite distance from the fixed point and the second close to the fixed point. We fix the initial value of ω at $\omega_0 = 1.5$ for the first set of experiments. Here, we vary P_m from 0.3 to 0.9 in the presence of a non-dimensionalized noise intensity of 3% in the system. A non-dimensionalized noise intensity of 3% means that the ratio of the RMS value of the angular velocity at the applied noise level is 3% to the RMS value of limit cycle oscillations in the absence of noise. For clarity, we denote the non-dimensional noise intensity as σ. As time evolves, the system parameter P_m is varied at predefined fixed rates as in Eq.4.2. We considered three different rates for our investigation as in the literature [139]. The three different rates considered for the study are $\mu_3 = 0.0001$, $\mu_2 = 0.0002$, and $\mu_1 = 0.0003$. The trajectories are plotted against the quasi-static bifurcation diagram to distinguish the effects of rate and noise. In order to plot the quasi-static bifurcation diagram, we vary the non-dimensional mechanical power, P_m both in the forward and reverse directions in a quasi-static manner as performed in Chapter 3. Fig.4.6(a) presents the transitions for the variation of mechanical power as a function of time, $\omega(t)$ of the canonical power system model for different rates of evolution of system parameters corresponding to $\omega_0 = 1.5$.

We notice an effect of the early crossing of the unstable manifold for higher rates of variation of mechanical power, irrespective of the initial P_{m0} value. Specifically, the point at which the system crosses the unstable manifold while nearing the Hopf point is dependent upon the rate at which P_m evolves. As a result, for systems with higher μ, the system crosses the unstable manifold and gets attracted to the stable limit cycle at a lower value of P_m.

Next, we repeat the experiment for the same rate of evolution of the system parameters as earlier, except for a different initial condition, when $\omega_0 = 0.1$. We observe that the dynamic trajectories, $\omega(t)$ crossover the unstable manifold without maintaining a specific trend with respect to rate in Fig.4.6(b), quite different from Fig.4.6(a). Then, to analyse the reason behind the different order of crossing the unstable manifold, we closely inspect the dynamic trajectories near the bifurcation point. We identify that the order in which the system crosses the unstable manifold is greatly determined by external stochastic excitation for the second set of initial conditions. Unlike the earlier case wherein the order of cross over is dependent

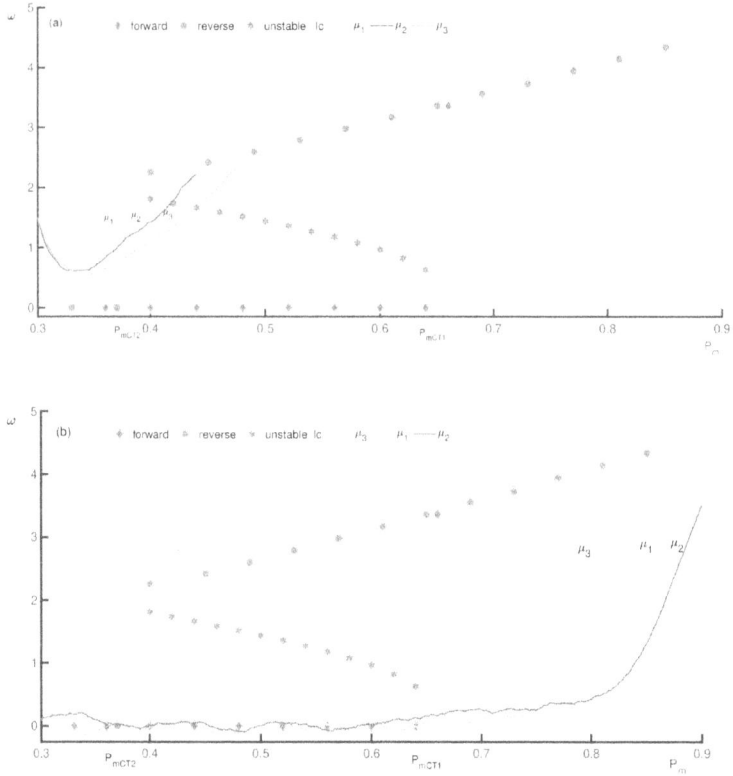

Figure 4.6: Dependence of transition characteristics on the initial conditions: Figure depicts the trajectory of angular velocity for two different initial conditions with respect to the quasi-static bifurcation diagram. (a) Away from the fixed point: the initial condition is specified as follows: $\omega_0 = 1.5$, $P_{m0} = 0.3$. The mechanical power is allowed to vary at three different rates, as in the order $\mu_3 < \mu_2 < \mu_1$ at a noise intensity of $\sigma = 3\%$ (b) Near the fixed point: : the initial condition is specified as follows: $\omega_0 = 0.1$, $P_{m0} = 0.3$. The mechanical power is allowed to vary at the same rate as in the earlier case with the same noise exposure. However, there is no definite trend on the transition characteristics, which is shown in the figure. The forward and backward paths of the quasi-static bifurcation diagram are indicated by blue and red coloured markers respectively.

upon respective μ, cross over depends upon the individual noise realisation for the operating condition close to the fixed point.

Additionally, we captured multiple realisations of the response while maintaining the noise intensity, the rate of variation, and the initial conditions constant to check the variability in stochastic realisation. As described earlier, we performed 50 simulations each for case1 and

67

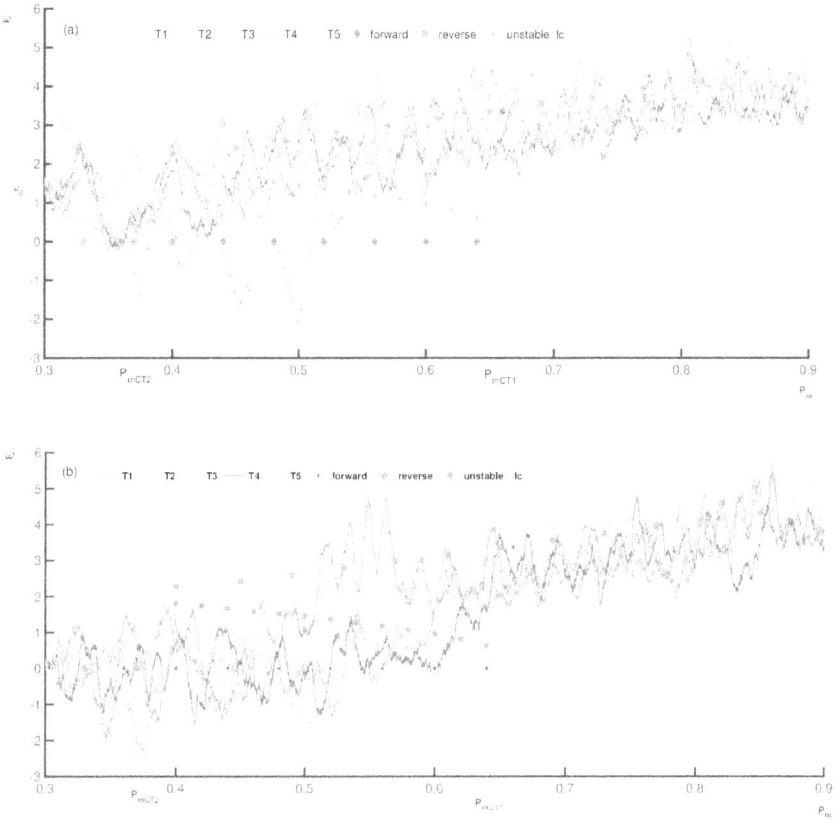

Figure 4.7: Variable transition characteristics with respect to initial conditions and fluctuations: (a) Away from the fixed point, $\omega_0 = 1.5$. (b) Near the fixed point, $\omega_0 = 0.1$. Figure (a) and (b) depicts the dynamic trajectories for the variation of mechanical power between P_{m0} and P_{mf} at a prefixed rate and noise intensity. Here, the dynamic trajectories are marked as T1-T5 with respect to the quasi-static bifurcation diagram for the mechanical power variation from $P_{m0} = 0.3$ to $P_{mf} = 0.9$. The forward and backward paths of the quasi-static bifurcation diagram are indicated by blue and red coloured markers respectively. The pink coloured hexagonal marks demarcates stable equilibrium point from the stable limit cycle.

2. We have plotted 5 realizations of the transitions corresponding to two different operating conditions, to ensure the clarity and to illustrate the spread in the point of transition, Case 1: $\mu = 0.0001$, $\omega_0 = 1.5$, $P_{mo} = 0.3$, $P_{mf} = 0.9$ and Case 2: $\mu = 0.0001$, $\omega_0 = 0.1$, $P_{mo} = 0.3$, $P_{mf} = 0.9$ when subjected to the same non-dimensional noise intensity of 3%. Fig.4.7(a),(b) demonstrate the multiple realisations of the transitions pertaining to case1 and 2. Our results are in qualitative agreement with the results observed by Unni *et al.* in thermoacoustic systems

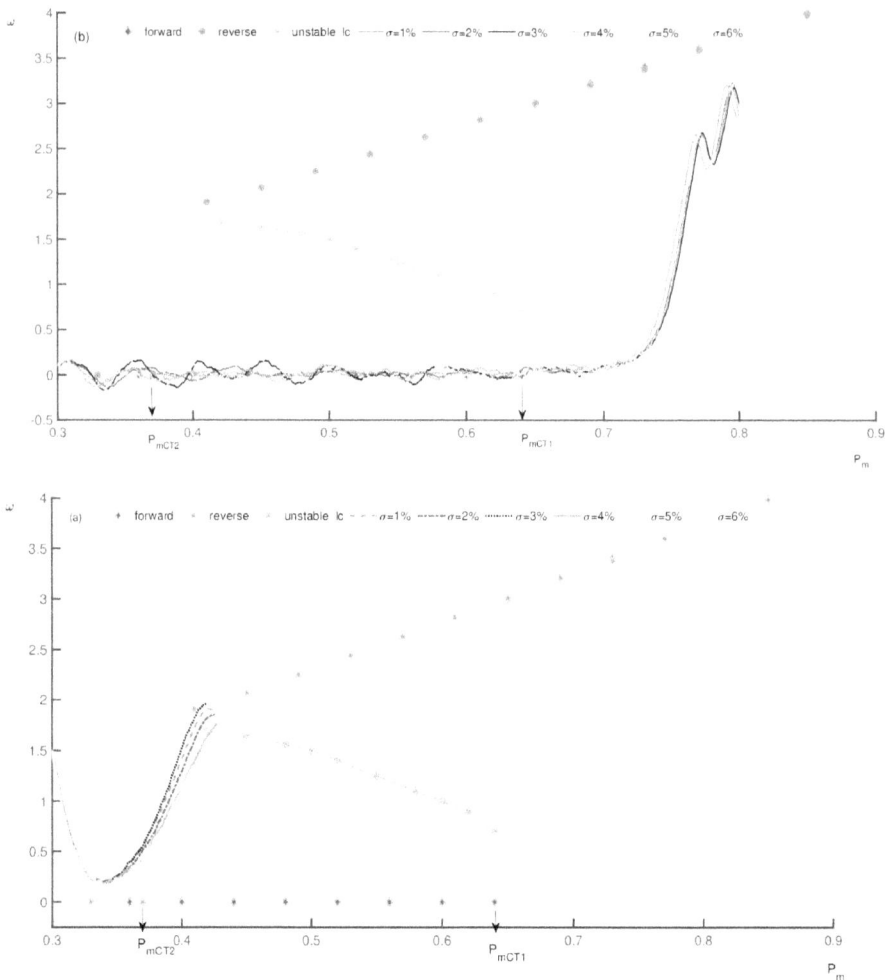

Figure 4.8: Influence of noise intensity on the transition characteristics:Figure depicts the angular velocity for two different initial conditions when exposed to varying noise intensities with respect to the quasi-static bifurcation diagram. (a) Away from the fixed point: the initial conditions are the same as in Fig.1 ($P_{m0} = 0.3, \omega_0 = 1.5$). The mechanical power is allowed to vary from 0.3 to 0.8 at a rate of $\mu = 0.0001$. The noise intensities considered in this experiment range from 1% to 6%. (b) Near the fixed point: the initial conditions are $\omega_0 = 0.1, P_{m0} = 0.3$ The average trajectory remains the same irrespective of the noise intensities considered.

[140].

The individual realisations of the trajectories exhibit substantial variability in the point of

transition to limit cycle oscillations in the presence of noise in Fig.4.7. We have computed the mean value of the transition point in both cases and computed the deviation in the transition point from the mean. This confirms the variability in the point of transition for each realisation. We observe the variability in the transition point for multiple realisations in Fig.4.7 with respect to Fig.4.6. However, the pronounced effect of variability can be observed in the plot, corresponding to the second case (Fig.4.7(b)), wherein the operating condition is close to the fixed point.

We infer that noise plays a crucial role in deciding the transition characteristic when the initial conditions are near the fixed point. It is identified that pre-bifurcation noise amplification is pronounced for slow transitions through the bifurcation point in systems with internal noise [139]. In our study, we show that when the magnitudes of the operating state variables are comparable with the magnitudes of noise intensities, the system becomes stochastic. This is dissimilar to the expected behaviour for the slow variation of system parameters through the bifurcation point. Thus, we observe that in the case of a stochastic bi-stable power system oscillator, there is an interplay between the noise and the rate, which determines the system response.

To identify the effects of noise intensity on the transition characteristics, we have repeated the experiments associated with Fig.4.7 for various noise intensities for a given rate of variation of mechanical power (Fig.4.8). We have considered 25 realisations of each noise intensity and computed the average trajectory to analyse the response in the presence of noise. We inferred that the dependence on initial conditions is preserved by inspecting the average trajectory of 25 realisations. As we have observed complete suppression of the bi-stable region for noise intensity, $\sigma = 8\%$, we consider noise intensities less than 8% to avoid the possibility of noise-induced transitions (NIT), which is beyond the scope of this work in our study. There is a qualitative similarity between our results on the shift in the point of transition in the presence of fluctuations and the change in the bifurcation point observed in the Henon map [141] and loss-driven CO_2 laser [142].

We find that stochastic power systems show significant randomness at the point of transition, resulting in a loss of 80% of the stability margin determined by the quasi-static bifurcation diagram in the studies that we conducted. This leads to an emergency state, which necessitates a dedicated control strategy [143], [144], [145], [146]. Therefore, it is extremely important to develop an effective control strategy for systems operating near to their physical limits, which is the subject of the next section.

4.3 Rate-driven control of power system operating near the physical limits

The transition characteristics of stochastic power systems presented in the previous section revealed that noise and the operating environments trigger transitions. Power systems usually operate near the stability limits for the best usage of the existing transmission assets [147]. Therefore, stochastic power systems operating near the stability limits are more susceptible to transitions observed in the previous section. We propose a control strategy for a stochastic power system operating close to the physical limits.

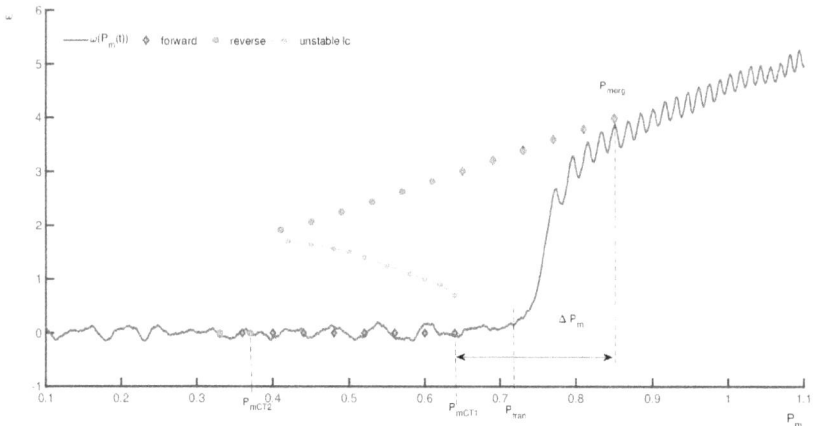

Figure 4.9: Margin for control of the system for variations of mechanical power as a function of time: The figure shows the transition to the oscillatory state for the mechanical power variation as a function of time and the merging point with respect to the quasi-static bifurcation diagram for $\sigma = 3\%$. The transition and merging point is respectively denoted as P_{tran}, and P_{merg} . The quasi-static bifurcation diagram is shown in blue and red cloured markers for the forward and reverse paths respectively. The window from the quasi-static bifurcation point, Pm_{CT1} to merging point, P_{merg} is marked as ΔP_m. The control parameter vartiation is performed at a rate of $\mu = 0.0001$.

We consider a power system model that operates beyond the quasi-static bifurcation point to resemble a physical system operating close to the limits. In our previous experiments, we noticed that the power system exhibits a delay in transition when the control parameter varies as a function of time [136]. Here, we notice a small window from the point of transition to the point at which the trajectory grazes the limit cycle oscillations. The point of transition and the point at which the trajectory grazes the limit cycle oscillations are respectively shown as P_{tran} and P_{merg} in Fig.4.9. Therefore, region of interest to develop a control scheme is from the bifurcation point, P_{mCT1} to the merging point, P_{merg}, denoted as ΔP_m in Fig.4.9. The time taken while traversing from P_{tran} to P_{merg} is dependent upon the rate at which the parameter is varied.

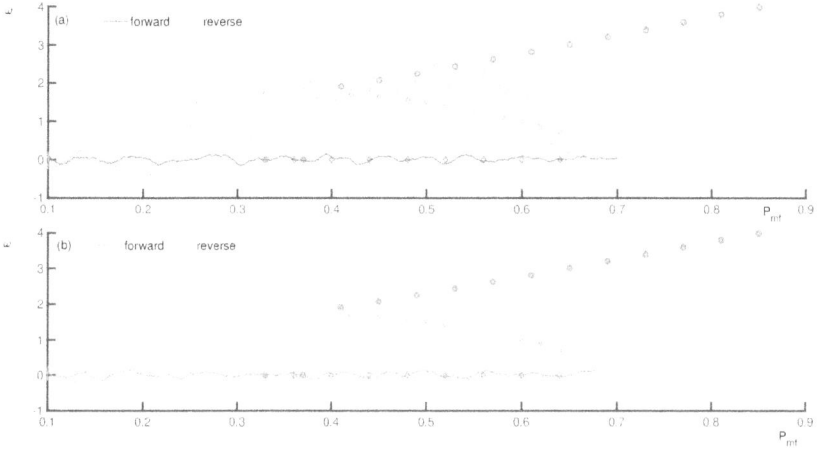

Figure 4.10: The rate-driven control strategy: We vary the mechanical power at $\mu = 0.0001$ from $P_{m0} = 0.1$ to P_{mf} and decrement from P_{mf} at the same rate for $\sigma = 3\%$. The response of the system for the increment and decrement of P_m are shown by blue and red coloured trajectories against the bifurcation diagram in Fig. 4.10. In Fig.4.10(a), $P_{mf} = 0.7$, where we can observe the transition to the stable oscillatory state, despite the decrement in P_m. Fig.4.10 (b) $P_{mf} = 0.68$ where the system regains the stable non-oscillatory state via the control strategy.

We vary the mechanical power from an initial value, P_{m0}, to a final value P_{mf} at a rate of $\mu = 0.0001$, the lowest rate considered for the experiment. Further, we decremented the mechanical power at the same rate as the increment to check whether the decrement was sufficient to bring the system back to the basin of attraction of the fixed point. Initially, we choose a P_{mf} value close to P_{mCT1} in our numerical experiment. We incremented P_{mf} gradually in steps of 0.01 to identify the highest P_m for which the reverse transition to the stable fixed point is possible for $\mu = 0.0001$. We observed that for P_{mf} beyond 0.68, the angular velocity crosses the unstable limit cycle and enters the basin of attraction of the limit cycle, even though P_m is decremented at a rate of 0.0001. Fig. 4.10(a),(b) depicts two representative cases from the numerical experiments conducted. In Fig. 4.10(a), the trajectory of ω crosses the unstable limit cycle, while we decrement the mechanical power for a P_{mf} value of 0.7. The crossover to the stable limit cycle is shown in green coloured trajectory in Fig. 4.10(a). In the latter case, ω returns to the basin of attraction of fixed point (Fig. 4.10(b)) when P_{mf} is decremented gradually from 0.68.

Next, we performed the numerical experiment to determine the maximum P_m, which allows the reversal of transition by incrementing μ values while decreasing the mechanical power for three different noise intensities. We focus on the region between the quasi-static bifurcation point(P_{mCT1}) to the merging point(P_{merg}) to determine the μ_{min} for the P_{mf}. The results in the calibration curve in Fig. 4.11(a) illustrate the minimum rate of the decrement of the mechanical power, μ_{min}, to perform a smooth reversal from the emergency state to the regime of safe return for three different noise intensities, $1\%, 3\%$, and 5% respectively. We can ensure stability by operating at mechanical power decrements between μ_{min} and the maximum limit imposed by the physical constraints.

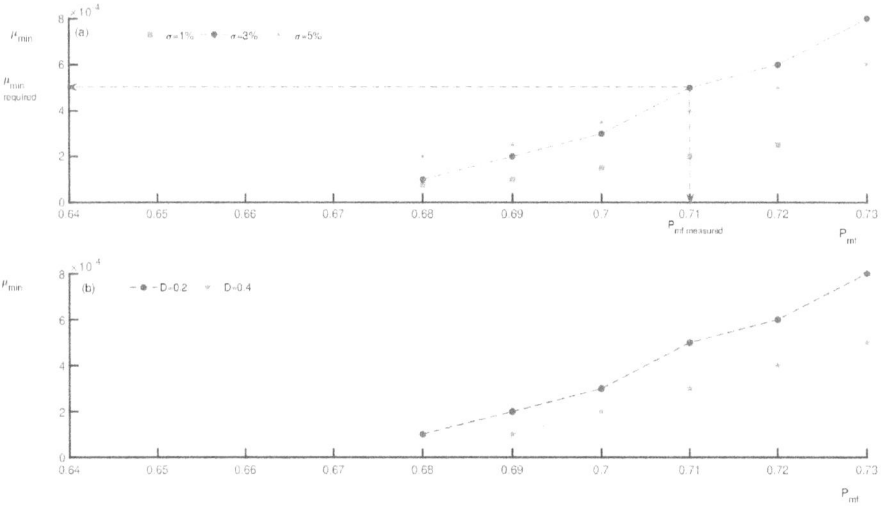

Figure 4.11: Calibration curve: is a plot which shows the rate of the decrement of mechanical power (μ) required to perform a smooth reverse transition of the power system from the emergency regime to the stable regime for $\sigma = 1\%, \sigma = 3\%,$and $\sigma = 5\%$ respectively. Fig.4.11(a), shows the calibration curve that illustrates the minimum rate required to revert the system back to stable non-oscillatory state for different P_{mf} values, when $D = 0.2$. Fig.4.11(b) shows the comparison of the minimum rate required for the decrement of mechanical power for two damping values. The figure depicts the reduction in the value of μ_{min} for an increase in damping performed at a noise intensity of 3%.

Further, we captured multiple realisations of the response for a predetermined μ_{min} and a given noise intensity(σ) to check the variability in the stochastic realisation. We performed the experiment for all the noise intensities considered in the calibration curve. By analysing the results, we observed that there exists variability in μ_{min} among multiple realisations of the same noise intensity. In an attempt to extend the P_{mf} value, we incremented the damping level of the system. We identified that it is possible to ensure a safe return to the non-oscillatory state with a lower rate of decrement of mechanical power by increasing the damping level of the system(Fig.4.11(b)). In Fig.4.11, the recommended region for operation pertaining to P_{mf} is above the blue and red trajectories for $D1$, $D2$ respectively. From the numerical experiments, we observe that if we perform the forward transition at $\mu = 0.0001$, there exists an upper limit for P_{mf}, beyond which the reverse transition to the stable fixed point will not be possible.

The proposed rate-driven control strategy for emergency control based on universal bifurcation regularity implies a practical application in different fields where a reverse transition from an unstable to a stable state is needed. The proposed rate-driven control strategy helps maintain the stability of engineering systems such as power systems [148], thermoacoustic systems [140], and in agroecological transitions [149], [150] when the systems are operating near the physical limits.

4.4 Concluding Remarks

In the present chapter, the effect of noise on the quasi-static stability characteristics of the power system model is presented initially. Our investigations revealed that the width of the bi-stable region gets reduced in the presence of noise. Further, we observed that there is a continuous transition between the two stable states in the presence of noise intensity beyond a threshold value, which leads to suppression of the bi-stale region. Then, the influence of initial conditions and stochasticity in triggering transitions in a non-autonomous power system is investigated. We observed that the system undergoes noise-induced transitions when the magnitudes of the noise intensity and state variable amplitudes are comparable. Furthermore, the effect of rate on the transition recedes in the presence of noise-induced transitions in power systems. This is dissimilar to the expected behaviour in systems which undergo a slow passage through the bifurcation. We established a boundary between the emergency state and the regime of the safe return that might help the operator to reschedule the generation dispatch.

Chapter 5

Early Warning measures for critical transitions in power systems

The previous chapters investigate the stability characteristics of canonical power systems with respect to the variation of system parameters and stochastic fluctuations. The electric power system is the central infrastructure for drinking water, telecommunications and banking services. Therefore, it is necessary to predict transitions in power systems and take ameliorative measures to reduce the impact of catastrophic transitions. In this chapter, we present the development of early warning indicators based on critical slowing down (CSD) for predicting oscillatory instability in power systems. Then, the robustness of these early warning measures for different rates of variation of system parameters and noise intensities is investigated. Finally, a deep learning algorithm is proposed to predict oscillatory instability that occurs in the power system model. Furthermore, the robustness of the proposed deep learning algorithm is analysed.

5.1 Prediction of instability mechanism in power systems

The major instability mechanisms in power systems are voltage instability and oscillatory instability [104]. There is sufficient literature that directly links voltage instability problems to saddle-node(SNB) and limit-induced bifurcation(LIB) [28], [151]. The computation of the minimum singular value of the Jacobian matrix and monitoring of the reactive power reserve of the generators predicts the point of collapse here [152]. Another significant instability, oscillatory instability, is associated with Hopf bifurcation [47].

Seydel *et al.* presented the determination of Hopf bifurcation(HB) based on the real part of critical eigenvalue for general dynamical systems in [153]. A model-based closed-loop monitoring system for the detection of HB in uncertain nonlinear systems was presented in [154]. The monitoring system uses the power spectrum of the measured output to identify oscillatory instability. However, the prediction is very close to the bifurcation point. Therefore, taking preventive actions is difficult, which hinders its practical applicability. Canizares *et al.* proposed a quasi-linear measure derived from the state matrix to make an early prediction of impending oscillatory instability [155]. The prediction of HB in power systems is based on the linearisation of the differential-algebraic equations (DAE) of the system [156]. The accuracy of the prediction is dependent upon the accuracy of the model. However, obtaining an accurate model of a system of several buses is difficult. Hence, research bifurcated to data-driven methods.

The large-scale implementation of wide area monitoring systems (WAMS) and phasor measurement units (PMU) facilitate identifying poorly damped modes and causal factors leading to instability by performing sensitivity analysis using the data [157]. Ghasemi *et al.* proposed a stochastic subspace identification method to extract the critical modes [156]. Liu *et al.* proposed a frequency domain approach based on singular value decomposition (SVD) for identifying the damping ratio, modal frequency, and mode shape of poorly damped oscillatory modes from the ambient PMU measurements [158]. The proposed quasi-linear index effectively predicts oscillatory instability during linear load increments. However, the index is not apt for systems that encounter discontinuities during load increments. Hence, this method is unsuitable for power systems with higher shares of renewable energy sources(RES).

Dakos *et al.* proposed a data-driven method to predict critical transitions based on the phenomenon of critical slowing down (CSD) exhibited by dynamical systems [79]. Afterwards, diverse dynamical systems ranging from physical systems to engineering systems have applied these early warning measures [79], [159]. Kuehn *et al.* observed that the system dynamics follow any of the 'normal forms' near the critical points, making these early warning measures suitable even for higher dimensional complex systems. Due to CSD, the subsequent state of the system resembles its previous states, which results in increasing 'memory' of the system, leading to an increasing trend in autocorrelation before a critical transition. Additionally, the rate at which the perturbations from an equilibrium state decay are reduced due to CSD, leading to an increasing trend in the variance as we approach the transition. Therefore, autocorrelation and variance were used as early warning measures for critical transitions in a physical system.

Cotilla *et al.* applied CSD- based measures to estimate the proximity of the operating point of a power system to that of critical transition [125]. The authors developed early warning signals for voltage collapse by analysing statistical parameters such as autocorrelation and variance in the power system data. The time series from a mathematical power system model and real-time synchronised PMU is used for developing CSD-based early warning indicators for voltage collapse. The main drawback of the proposed statistical indicator was that it could not indicate whether the operating trajectory would lead to instability with certainty.

Ghanavati *et al.* investigated the usefulness of autocorrelation as a precursive measure for SNB in power systems [82]. Three mathematical models of power systems were used for the investigation. The authors established that the CSD-based EWS appears very close to the point of transition, making it less effective as a precursor. The authors also noticed that the variance in voltage magnitude serves as a better indicator than autocorrelation. Further, the authors proposed a semi-analytic method for the computation of autocorrelation and variance of voltages and currents in a power system model [20]. However, it is reported that systems fail to show CSD-based EWS before a CT when the perturbations are not in the direction of the dominant eigenvector across which the system destabilises [86]. Subsequently, in another work, Ghanavati *et al.* illustrated the conditions under which the statistical indicators will provide reliable early warning of instability in power systems [20].

Zheng *et al.* proposed a monitoring and prediction framework to address voltage instability through SNB in a large-scale power system [88]. The authors noticed the monotonic increase in the variance of the state variable near the point of collapse. The authors also proposed a scaling law between the bifurcation parameter and variance to determine the transition point by an approximate linearisation method. Contrary to the above observations, silent transition to the alternate state and partial failure of these measures are also reported in the literature [85], [86]. Ke'fi *et al.* observed the presence of slowing down based-EWS before non-catastrophic transitions as well [87]. These investigations offer a cautionary note to the applicability of CSD-based EWS without analysing the robustness of these measures.

The development of CSD-based early warning indicators that undergo rate-dependent transitions is unexplored in power systems . Further, the robustness of these early warning measures for different rates of variation of the control parameter is not investigated in any dynamical system to the best of our knowledge. Additionally, the effectiveness of these measures in the presence of various noise intensities is not studied. Hence, the following section covers the development of CSD-based EWS for a stochastic power system that undergoes rate-dependent variations of system parameters, and varying intensities of AWGN. Further, the robustness of these early warning signals to predict oscillatory instability in power systems is investigated.

5.2 Early warning measures for oscillatory instability in power systems

This section aims to develop early warning signals to forecast the event of oscillatory instability in the power system model. Towards the same, we varied the control parameter at different rates as a function of time. We captured the time series that depicts the transition from a non-oscillatory state to an oscillatory state. Initially, the mechanical power is incremented at a rate of 0.0001/time step, and the associated angular velocity is recorded. The time series of angular velocity and the associated P_m variation is shown in Fig. 5.1.

We can observe that the time series depicts a sudden transition to oscillatory instability upon crossing the mechanical power value at $t = 136$. (Fig.5.2(a)). The P_m value at this point is 0.685. From our earlier investigations (illustrated in Chapter 3), we observed that the quasi-static Hopf point is at $P_m = 0.64$. The rate-dependent delay described in the same chapter is the reason for the delayed transition to oscillatory instability. We use the time series up to $t = 118.2$ for developing early warning signals. This time series contains 87.5% of the data ahead of the point of transition. This also corresponds to 93.5% of data points prior to the quasi-static Hopf point. We are restricting the time series up to this value because we intend to develop an early warning measure before crossing the Hopf point for taking an effective

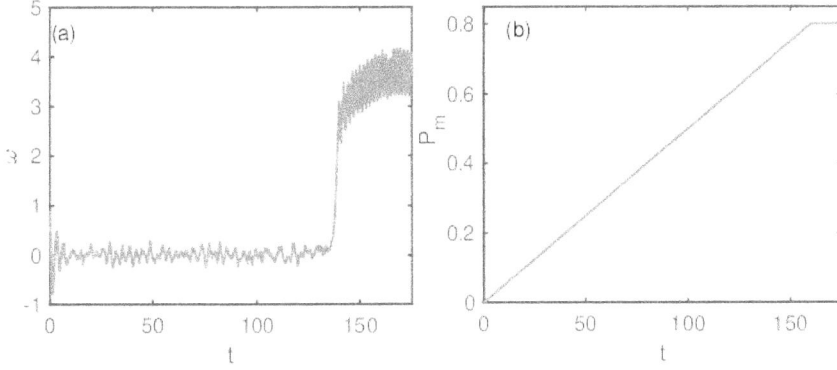

Figure 5.1: Figure depicting the time series of ω and variation of P_m. (a) The response angular velocity for mechanical power variation as a function of time at noise intensity of $\sigma = 2\%$, keeping all other system parameters the same as in previous chapters. (b)The variation of mechanical power at a rate of 0.0001/ time step from 0 to 0.8

preventive measure.

We can see a significant increase in the value of autocorrelation prior to the transition, serving as a potential early warning signal (Fig.5.2(b)). A similar incrementing trend in the value of standard deviation can be seen before the transition(Fig.5.2(c)). We also observe a decrementing trend in the return rate before the sudden transition, which also serves as an early warning(Fig.5.2(d)) [77]. The slope of the linear fit to the autocorrelation, variance and return rate is computed, which serves as the trend.

We observed that the incrementing trend in standard deviation to be more pronounced compared to autocorrelation. It is reported in the literature that autocorrelation may turn out to be a less effective precursor in the presence of fluctuations [160]. Therefore, we performed multiple realisations of the transition to identify whether autocorrelation can be used as a precursor for transitions in power systems. We simulated 100 realisations of our stochastic SMIB model with $\sigma = 2\%$ for the same. Further, we incorporated AWGN of different intensities to mimic a physical power system operating in a stochastic environment. The remaining noise intensities considered for the experiment are 4% and 6%. The procedure for adding noise is elaborated in Chapter 2. The performance evaluation measures for rate-dependent variation of mechanical power (0.0001/ time step) in the presence of three noise intensity levels considered are summarised in Table. 5.1.

Further, we repeated the experiment for two different linear rates of variation of mechanical

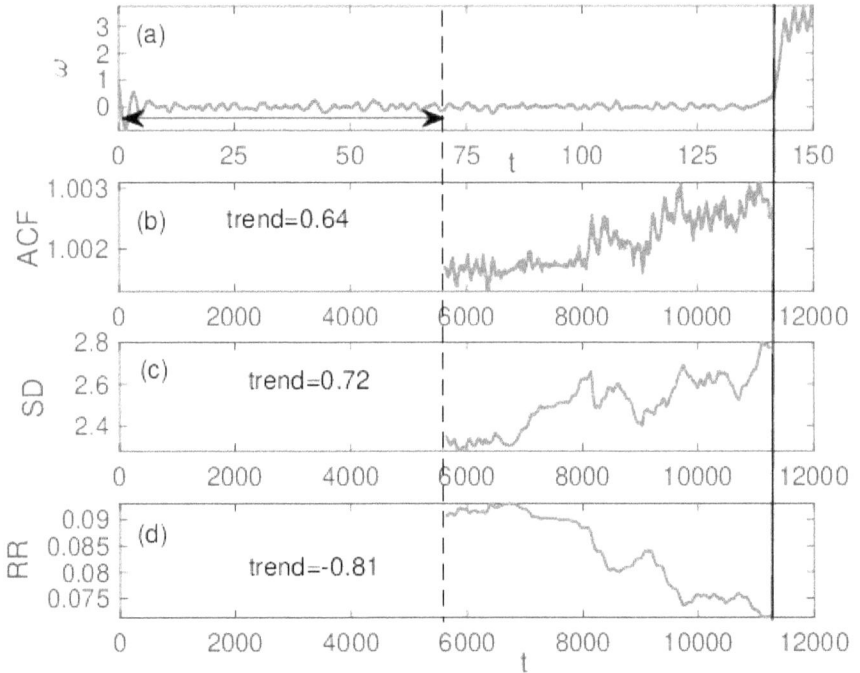

Figure 5.2: Early warning indicators of critical transition associated with oscillatory instability derived from time series of angular velocity (a)Time series illustrating the shift from the non-oscillatory to the oscillatory state for the continuous variation of mechanical power from 0 to 0.8 at a rate of 0.0001(b)the autocorrelation at lag–1(c) the standard deviation(d) the return rate of the system upon nearing a critical transition. The length of the moving window is fixed as half the size of the time series. A black horizontal arrow (Fig.5.2(a)) represents the length of the moving window. A solid vertical black line segment demarcates the time series and the data utilized for generating early warning indicators. The black dotted line indicates the time stamp from which early warning signals are developed. We observe an explicit increase in lag–1 autocorrelation and standard deviation, whereas a decrease is observed in the return rate of the system well before the transition. The data is generated in the presence of AWGN of $\sigma = 2\%$ to the power system, maintaining all other system parameters constant. We have used the early warning toolbox(EWT) available in R to compute the early warning indicators.

power to investigate the influence of rate on the CSD-based EWS. We restrict the rate variation until $r = 0.001$, corresponding to the early take-off observed in chapter 3. The rates considered are $r_2 = 0.0005$, $r_3 = 0.001$. The performance measures obtained for different rates and noise intensities are presented in Table.5.1.

We inferred that CSD-based early warning signals exhibit false positive(FP) and false negative(FN) results, making predicting transitions in power systems using those not robust. The variability in the stochastic realization are the reason behind false positive and false negative

Table 5.1: Scores of the performance measures of CSD based EWS for different levels of noises and mechanical power variation rates

Performance scores	μ_1			μ_2			μ_3		
	$\sigma = 2\%$	$\sigma = 4\%$	$\sigma = 6\%$	$\sigma = 2\%$	$\sigma = 4\%$	$\sigma = 6\%$	$\sigma = 2\%$	$\sigma = 4\%$	$\sigma = 6\%$
Accuracy	0.64	0.57	0.48	0.55	0.51	0.47	0.45	0.43	0.42
Precision	0.70	0.60	0.48	0.58	0.51	0.46	0.44	0.43	0.42
Recall	0.47	0.43	0.30	0.35	0.35	0.32	0.44	0.38	0.37
F1 score	0.57	0.50	0.37	0.44	0.41	0.38	0.44	0.40	0.40

results, reported in the thesis. Additionally, the performance scores exhibited a decrementing trend for higher mechanical power variation rates (Table. 5.1). Hence, the CSD-based EWS is found not robust in predicting oscillatory instability in stochastic power systems. This prompted us to develop a robust early warning indicator for predicting oscillatory instability in the power system model. The ubiquitous influence of artificial intelligence(AI) in predicting transitions in physical [161] and engineering systems [162] from the time series data motivated us to develop an AI framework suitable to power systems as well. In the following section, the AI-based prediction framework is presented.

5.3 Deep learning algorithm for generating early warning signals for impending transitions in power systems

The state-of-the-art literature on the efficacy of supervised deep learning algorithms demonstrates the ability to classify time series based on salient features in the data [163]. The effective functioning of the supervised deep learning algorithm to classify time series when trained with abundant data is reported [164]. In this background, the possibility of classification with a DL algorithm by studying CSD and other elusive features present in the data is explored in the subsequent section. Therefore, it can be surmised that DL algorithms can be employed to identify transitions in the time series. Nonetheless, supervised DL algorithms need thousands of time series data to make a correct classification [165].

The efficacy of feature-based models [166], ensemble models [167] and deep learning models [168] for the time series classification have been reported in the literature. Recurrent neural networks (RNN) effectively use temporal information for classification and prediction [169].

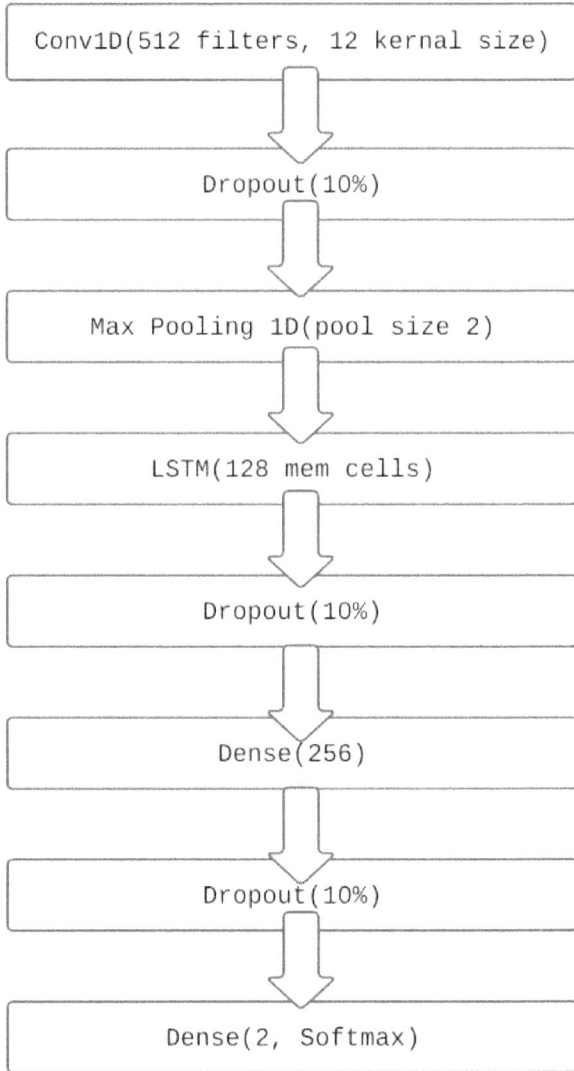

Figure 5.3: Architecture of the proposed CNN-LSTM model

The special architecture of the LSTM for sequence pattern information is also reported [170]. Convolutional neural networks(CNN) are reported to filter out noise in the data and extract salient features [170]. The development of ensemble deep learning algorithms for providing EWS of tipping points in real-world systems is reported in [120]. Therefore, we developed an ensemble architecture of CNN and LSTM for providing early warning signals for oscillatory instability in power systems.

The proposed CNN-LSTM is implemented using Keras, an open-source neural network library in Python. The architecture of the proposed CNN-LSTM model is shown in Fig. 5.3.

In the proposed CNN-LSTM architecture, two different neural network layers are sandwiched. The 1st CNN layer reads in subsequences of the time series and extracts features present in the subsequences [120]. This information is then passed to the LSTM layer, which interprets those features. The LSTM layer is implemented to learn the long-term dependencies in the data set. Finally, the dense layer processes the output of LSTM and learns features from all combinations of the previous layer. As a result, this approach excels at sequence prediction [164], [165].

We generated data sets from our power system model for three mechanical power variation rates considered earlier. The initial mechanical power variation considered for the experiment was 0.0001/ timestep. Then the data set is divided into training and testing to ensure the validity of the proposed model. We have used a data split of 0.7425/ 0.075/0.25 for the training, validation and testing, respectively. The grid search algorithm is used for the tuning of hyperparameters. The parameters considered for the grid search are shown in Table. 5.2.

Table 5.2: The set of hyperparameters used for training the model

Parameter	Set of hyperparameters
Hidden layer neurons	(32, 64, 128)
Activation function	(Softmax, ReLU)
solver	(SGD, Adam)
learning rate	(0.001, 0.0001, 0.01)

The optimal model obtained after the grid search is shown in Table.5.3. Afterwards, we trained the CNN-LSTM architecture with the training data to forecast whether the time series undergoes a critical transition with the optimal model for 100 epochs. We have given early stopping of the training by monitoring the loss. We used 87.5% of the data from the time series before the quasi-static bifurcation point for training as in the previous method. The performance analysis of the deep learning algorithm in the presence of three different noise

Table 5.3: The hyperparameters of the proposed architecture

Sl. No	Data Specifications	
1	Training data	2400x2352x1
2	Testing data	600x2352x1
	Model Specifications	
3	Type of convolution	One dimensional
4	No. of filters	512 with Kernal size of 12
5	Activation at CNN and Dense layer	Sigmoid
6	Activation at LSTM	ReLU
7	Optimizer	Adam
8	No. of hidden layers	128 at LSTM
9	No. of training iterations(epochs)	100
10	Batch size	32
11	Loss function	binary cross entropy
12	Learning rate	0.0001

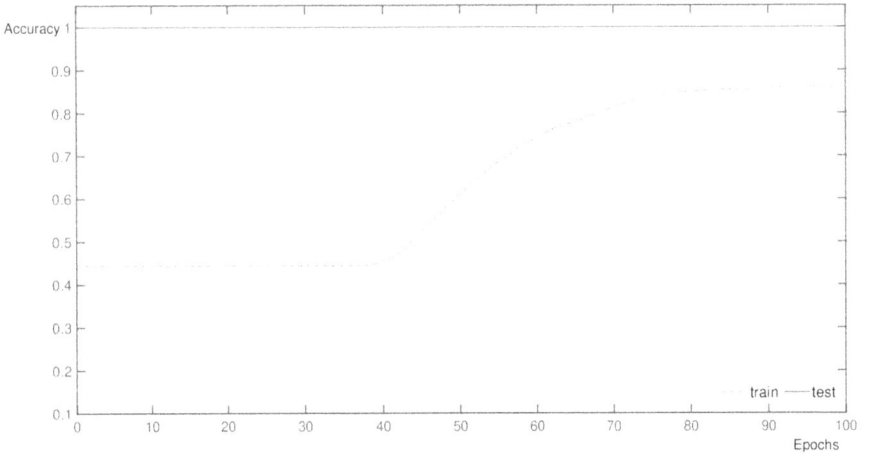

Figure 5.4: Training and testing accuracy of the proposed CNN-LSTM model with respect to epochs

intensities is presented in Table.5.4 for the validation of the proposed architecture. Then, we varied the mechanical power at parameter at $\mu_2 = 0.0005$/timestep and $\mu_3 = 0.001$/timestep. The rate μ_3 corresponds to the early take-off illustrated in Chapter 3.

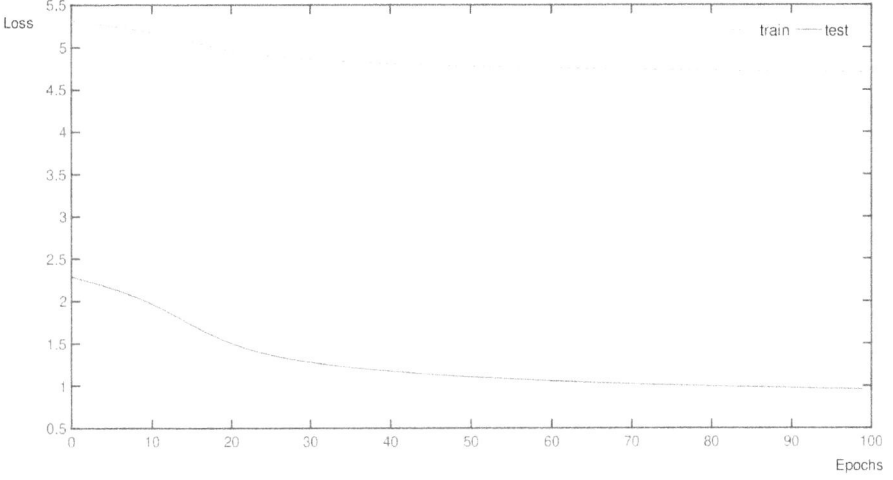

Figure 5.5: Training and testing loss of the proposed CNN-LSTM model with respect to epochs

Table 5.4: Scores of the performance measures of the DL algorithm for different levels of noises and mechanical power variation rates

Performance scores	μ_1			μ_2			μ_3		
	$\sigma = 2\%$	$\sigma = 4\%$	$\sigma = 6\%$	$\sigma = 2\%$	$\sigma = 4\%$	$\sigma = 6\%$	$\sigma = 2\%$	$\sigma = 4\%$	$\sigma = 6\%$
Accuracy	1	0.97	0.96	0.90	0.85	0.84	0.75	0.72	0.70
Precision	1	1	1	0.91	0.87	0.87	0.7130	0.6419	0.6908
Recall	1	0.95	0.93	0.91	0.85	0.85	0.7533	0.7297	0.7087
F1 score	1	0.98	0.97	0.91	0.86	0.86	0.7280	0.6766	0.6959

Accuracy and loss are the measures used to validate the proposed architecture. The model accuracy and loss are shown in Fig.5.4, and 5.5, respectively. The performance scores of the model for the rates are presented in Table. 5.4. We have used repetitive padding for the higher rates of variation of control parameters($\mu = 0.0005$, $\mu = 0.001$) [171].

5.4   Concluding Remarks

Multiple realisations of the transitions exhibited contradictory results for CSD-based EWS in the presence of stochasticity. Therefore, CSD-based early warning signals are not robust enough to forecast oscillatory instability in stochastic non-autonomous power systems. The robustness of CSD-based EWS is presented with the help of performance scores such as accuracy, precision, recall, F-1 score. We have considered different noise intensities and rates for mechanical parameter variation for the analysis. Further, we presented the CNN-LSTM algorithm for forecasting oscillatory instability in the power system model. Our results show that the proposed ensemble DL architecture is robust in identifying the transitions, even in the case of advanced transition. The findings from this study are crucial since there exists literature on the failure of CSD-based EWS in many engineering systems. The study suggests that in some situations, it is possible to prevent the occurrence of catastrophic transitions by a combination of CSD-based EWS along with deep learning architecture.

www.ingramcontent.com/pod-product-compliance
Lightning Source LLC
Chambersburg PA
CBHW071500210326
41597CB00018B/2635